THE POSTMODERN AESTHETICS DESIGN

深圳视界文化传播有限公司 编

国际创新居住空间赏析

APPRECIATION OF INTERNATIONAL INNOVATIVE LIVING SPACE

中国林业出版社
China Forestry Publishing House

PREFACE
序言一

BRING A DREAM TO ENCOUNTER MORE CREATIVITY ON THE JOURNEY
带上梦想让设计去旅行

全球化已经颠覆了传统的沟通守则：社交网络与互联网让我们可以如此轻松地交流，大到国家或地区的科技专长，小到个人的创新点子，都可以通过一个简单的“点击”跨越国界，发光于世界的每个角落。

建筑和室内设计正在持续发热，越来越多的人更重视自己的家居设计，大众充满了对潮流和新鲜事物的好奇。很多艺术家、建筑师和装饰师们的作品被跨越国界传播，他们的才华吸引了来自世界各地人们的关注。因此，其中有的人被贴上了某种风格的标签，也许这个标签让他们成功了，但某种程度上，却固化了思维，限制了他们的创造力。

室内设计一直是由大流行趋势而定，就我而言，希望保留自己风格的自由，不去迎合任何一种具体的设计风格。我期待每个新的作品，它们都是一次全新的会晤和交流，可以任由感觉去引导与寻找。所以一直以来，我非常欣赏“艺术之家”的概念：我们从来不专门为了一个客人单独设计一个场所，而是建议直接向买主们、物业和公寓出售我们的服务，根据我们的标准进行全面的装修，并且最终保留产权。我渴望人们能通过作品中所讲述的故事来认可我的设计，从视觉设计的修辞手法中去认识我的风格！设计需要有创意，从而创造一些意想不到的东西，避免常有的惯性思维。融入古典美和现代艺术，更新换代，完成一个独一无二的设计，给设计第二次生命与活力。

历史的影响在后现代设计中起着重要的作用，有折衷或渐进的转变。后现代主义的关键原则是复杂性和矛盾性，现代设计所提出的乌托邦完美主义已经被建构主义和美学所取代。后现代设计更强调艺术性，会采用比较夸张的手法，选择一些非常规的创意。其中生动而矛盾的色彩，别致的形状和狂野的图案，孟菲斯的家具等，一系列充满创造力的想法为空间注入鲜活的生命，在作品传达的意象里，会有“一见钟情”的欢欣。

我作为一个土生土长的巴黎人，巴黎厚重的历史使我不得不对其美轮美奂的建筑遗产肃然起敬。穿梭在这些奥斯曼建筑中，那些高高的天花板，那些地板，和那些漂亮的大理石壁炉等，都使我无法忽略掉这个城市的过往。但是她又必须随着时间不断地演变发展，因此我们需要给她注入新鲜的活力，让她能一直闪耀下去。就如我们做设计一样，需要不断地注入新的创意与想法，让设计充满未知的美感！

Globalization has overturned the traditional communication rules: the social networks and the Internet allow us to communicate with each other easily. Not only the expertise of a nation or a district but also the personal innovative ideas can cross the national borders to influence around the world by a simple "click" action.

The architectural design and interior design are being popular. More and more people pay more attention to the interior design of their home, and the public is full of curiosity about the trends and the new things. The works of many artists, architects and designers had been spread across the national borders to the world, and their talents had attracted the attentions of the people from all over the world. Therefore, some of them are labeled with a certain style that may have made them successful. But, to a certain extent, it solidified their minds and froze their creativity.

The interior design has always been determined by trends. For my part, I want to keep the freedom of my style which not cater to any specific design styles. I look forward to every new project, because it is the result through the new meeting and communication, and you can be guided and find it by the inspiration that it gives. Therefore, for a long time, I really appreciate the concept of "art homes": instead of designing a place for a single client, we suggest that buyers, properties and apartments purchase our services directly, totally renovated and decorated the projects according to our standards and finally retain their property rights. I long for people to accept my design by the stories of the projects and to recognize my style by the rhetoric of the visual design. The design needs creativity, so that I create something unexpected to avoid the usual inertial thinking. I combine the beauty of the past and the contemporary art and renew them to finish a unique design and give the design a new life and vitality.

The influences of the history play an important role in the post-modern design with the eclectic and progressive twist. The key principles of the postmodernism are complexity and contradiction. And the perfect Utopian, as proposed by the modern design, has been replaced by the constructivism and aesthetics. The post-modern design puts emphasis on artistry, adopts some exaggerated approaches and selects some unconventional ideas. The vivid yet contradictory colors, the unique forms, the wild patterns and the Memphis furniture, etc., a series of these creative ideas inject the living life into space. In the image conveyed by the project, there will exist the joy of "love at the first sight".

As a native Parisian, I respect and love her profound history and magnificent architectural heritage. Walking through these Ottoman architectures, their high ceilings, floors and beautiful marble fireplaces make me cannot ignore the past of this city. But she will continue to evolve over time, and there is a need to provide a new impetus to her for continuing to shine. Just like we do the design, we need to continuously inject new creativity and ideas to make the design full of the unknown aesthetics!

GÉRARD FAIVRE
GÉRARD FAIVRE PARIS

PREFACE
序言二

BRING FORTH THE NEW FROM THE OLD, PURSUE THE SURPRISES OF SPACE
推陈致新，追寻空间的惊喜

2018新年伊始，受邀为本书《后现代美学设计》写序，深感荣幸。回顾这些年的设计历程，感慨良多。跨界室内设计5年，每一年都有不同的悸动与欣喜，无论是在作品案例的表现手法上，还是在与行业内外的分享交流平台上，我都会用自己的见解、心得去体会每一个当下的风格走向以及未来的流行趋势，同时提炼出了属于自己的百态人生。

我是从平面广告转到室内设计的，在创意概念、色彩搭配、比例构造等方面以广告人的手法为特色，而介入室内更重要的是表现创造性以及空间的合理布局，因而我在设计时也会考虑得更为全面，并不断挑战新的事物。我认为创新与时尚是一种生命的延续，那种不安于现状的时代特性，是让我不断推陈出新、追求新鲜事物的动力。这些年我的案例没有特定的风格界限，更希望用天马行空的创想去落地项目中的每一处细节，不拘泥于传统的设计思维方式，而是用全新的手法去重组融合，时不时在现代的手法中穿插古典的元素，让空间中既有时尚个性的当代感，又有古典历史的延展性，同时点缀一些原创产品设计，使整个案例更有包容性和穿透力，也更富有生命气息，这也是我认为的“亦此亦彼”的设计特色。

“人事有代谢，往来成古今”，在今后的一段时间，我希望区别于标准化的现代主义空间，多以人性经验为导向，强调时空的统一性、延续性以及历史的互渗性，散漫与自由化的后现代风格或类后现代风格方兴未艾，这种趋势还将不断地日趋完善下去。它不能仅仅以所看到的视觉形象来评价，还需要我们透过形象从设计思想角度分析。而对于这样的空间氛围营造，我们需要从不同的专业领域出发，去融入这个视觉呈现。比如我们可以去看一些奢侈品牌的主题秀，也可以关注每一年的国际时装潮流发布会，更可以是一些大型的橱窗系列布展，每一个拥有创意的地方都会是我们汲取经验和获取灵感的所在。通过种种方式，我们再进行一次又一次的头脑风暴，接受不同思潮的洗礼，融合自己对这些创意的奇妙感受，让自己在思想激荡中带来与众不同的感观体验，进而做出属于自己的独特创意设计。

It is a great honor to be invited to write a preface to this book,*The Postmodern Aesthetics Design*, at the beginning of 2018. Looking back over the years of the design experiences, I had a lot of feelings. Changed my job to interior design for 5 years, I had different throbs and joys. Whether in the case of the expression of the work, or in the sharing platform with the design circle, I not only use my views and experiences to understand each current style and future trend but also live my life.

I have changed my profession from print advertising to interior design. In terms of creative concepts, color collocations, proportional constructions, etc., I take the technique of an advertiser as a feature, while the more important things for interior design are the creative expression and the rational layout of space. Therefore, in the process of design, I will consider it more comprehensive and constantly challenge the new things. I believe that innovation and fashion are the continuations of life, and the characteristic of Era that is not content with the current status, is the driving force for me to constantly bring forth the new from the old and pursue the new things. In these years, my works have not been restricted by the certain styles, but I hope that my works achieve every detail of the project with the imaginative creativity, are unrestricted by the thinking mode of traditional design, use a new approach to reorganize and merge, and intersperse classical elements from time to time, so that space has both a contemporary sense of fashion and personality and the extensibility of classical history. At the same time, I will use some original products as the embellishments to make the whole case not only more inclusive and transparent but also rich in the flavor of life. These are the "unify of the opposite" of design features.

"While worldly matters take their turn; time comes, time goes, thus there are old and modern days." For the next few years, I hope the modern-style space distinguished from the standardization can take the human experience as guidance, emphasize on the unity and continuity of the time and space and the interpenetration of history. The discursive and liberalized postmodern style or the similar postmodern style is in the ascendant, and this trend will be improved gradually. It can not only be judged by the visual image we have seen, but also need us to analyze it from the perspective of a design idea. And for creating this kind of space atmosphere, we need to integrate this visual presentation from different professional fields. For example, we can go to some thematic shows of a luxury brand, also we can pay attention to the international fashion press conference and some large window exhibitions. Every place where has creative ideas is where we can learn from the experiences and gain the inspirations. Through a variety of ways, we have been brainstorming again and again, receive the baptism of different trends of thoughts, integrate my wonderful feelings of this creativity, and bring a different visual experience in the exchange of thoughts to myself, so that I can do my unique creative design.

GE XIAOBIAO
葛晓彪
JYM DESIGN
金元门设计

CONTENTS
目录

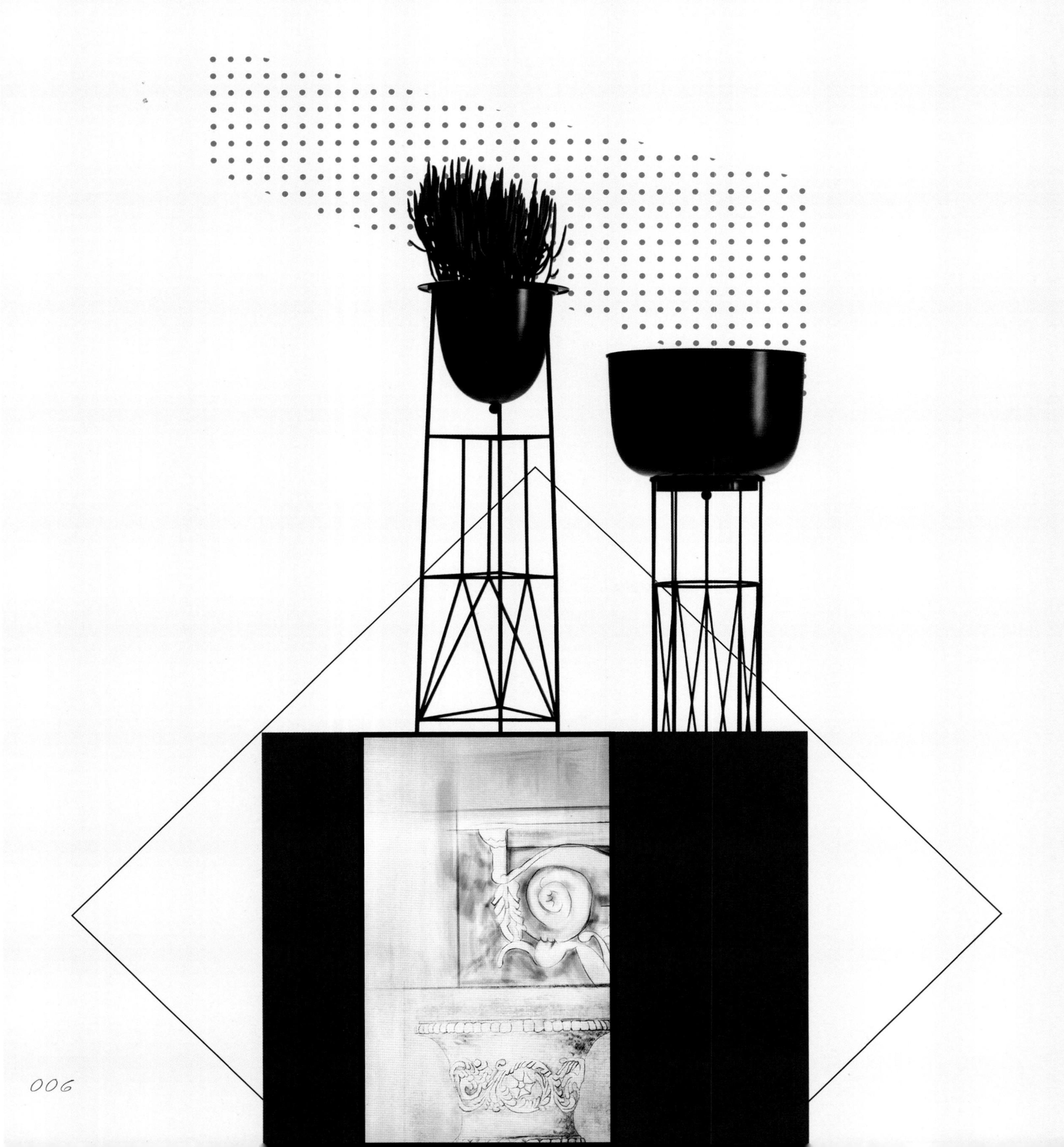

THE REBIRTH OF THE ANCIENT ARCHITECTURE

Project Name | Avenue Marceau

Design Company | Art homes by Gérard Faivre Paris

Designer | Gérard Faivre

Project Location | Avenue Marceau, Paris

Area | Around 272 m²

Photographer | Mathieu Fiol

古建筑的今世重生

Gérard Faivre Paris, the creator of the 'Art Homes' concept, has revolutionised Parisian luxury real estate by offering fully renovated, ready-to-live-in, prestigious apartments for sale. These apartments never lived in before all boast a superior interior design that has transformed them into genuine works of art, with each place equipped with a private home concierge service, to make new residents feel at home right away.

The 8th arrondissement in Paris on the right bank of the Seine attracts people from all over the world. Areas such as the Champs-Elysées, Avenue Montaigne, Avenue Matignon and Avenue George V not only are big draws for visitors to Paris, but also make up a popular affluent residential district in the city.

This apartment designed by Gérard Faivre is situated right in the heart of Paris' "Golden Triangle", the most prestigious shopping area of this arrondissement, on Avenue Marceau, just 100 metres from Place de l'Etoile and the Champs Elysées.

Gérard Fivre Paris，“艺术之家”概念的创造者，通过提供全新的、现成的、有声望的销售公寓，革新了巴黎奢华的房地产行业。这些从来没有入住过的公寓都拥有优越的室内空间布局，设计师将其改造成为真正的艺术作品，给每个地方都配备私人礼宾服务，力图让新住客有家的感觉。

在巴黎第八区的塞纳河畔吸引着世界各地的游客们。香榭丽舍大街、蒙田大道、马蒂尼翁大道和乔治五世大道等地区不仅吸引了大量游客到巴黎，还构成了这个城市一个备受欢迎的富裕住宅区。

这套公寓是由Gérard Faivre设计的，位于巴黎“金三角”中心，是玛索大道上最著名的购物区，距离星形广场和香榭丽舍大街只有100米。

KASPER

The 272 square-metre apartment is on the third floor of a traditional stone Haussmannian building which dates back to 1880. It offers stunning views of Avenue Marceau and the Arc de Triomphe, making this bright and luxurious apartment an ideal pied-à-terre for a foreign clientele to enjoy quintessential Paris scenes.

For this project, Gérard Faivre collaborated closely with artists, filling the apartment with carefully selected creations to accentuate the uniqueness of the home. Paintings, photographs and sculptures were painstakingly selected to be in perfect harmony with the atmosphere Gérard Faivre wanted to create.

这套272m²的公寓位于一栋建成于1880年传统的豪斯曼尼亚石楼的三楼。它有着令人叹为观止的视野，可以直接欣赏从马尔索大道到凯旋门的景观，这使这间明亮豪华的公寓成为外国客户欣赏巴黎美景的理想场所。

在这个项目中，Gérard Faivre与艺术家们紧密合作，精心挑选的作品摆满了整个公寓，以强调它的独特性。挂画、照片和雕塑都是经过精心挑选出来的，与Gérard Faivre想要创造的气氛完美地融合在一起。

THE EXTRAORDINARY TENSION, BLACK AND WHITE IN CONTROL OF MODERN ARTISTIC LIFE

非凡张力，黑白掌控摩登艺术生活

Project Name | Todeschini Pavilion

Design Company | Studio Guilherme Torres

Designer | Studio Guilherme Torres

Project Location | São Paulo, Brazil

Area | Around 400 m²

Photographer | MCA Estudio

You have the nature of self-conceit, but you fall in love with the low-key style. You have an outstanding appearance, but you are far away from the public. Under your every gorgeous color, you are showing the varied and unusual life. This is the charm that you give the home space.

Black and white as a low-key, classical collocation with a light luxury style, allocating with the rational fabric with a geometric pattern, the echoing geometric lines, and the tranquil wood, these simple and introverted colors exude an elegant and intellectual temperament, which is intoxicating. With the contrast of rigid geometric pattern, the monochromatic bright white fabric and the large hanging crystal chandelier highlight the simple theme, and present the space beauty of plain and elegance.

你有自负的天性，但你爱上了低调；你有出色的颜值，但你又远离江湖。在你每一个华丽的色调之下，演绎着形形色色不同寻常的人生，这就是你赋予家居空间的魅力！

黑白作为低调轻奢范儿的经典搭配，携手代表着理性的几何图案布艺，几何线条的呼应，加入宁静的木质，其简约内敛的色调彰显着优雅知性气质，令人心醉沉迷。单色亮白布艺与垂落的大型水晶吊灯，在刚性几何图案的对比之下，突出简约主题，更加呈现出空间的素雅之美。

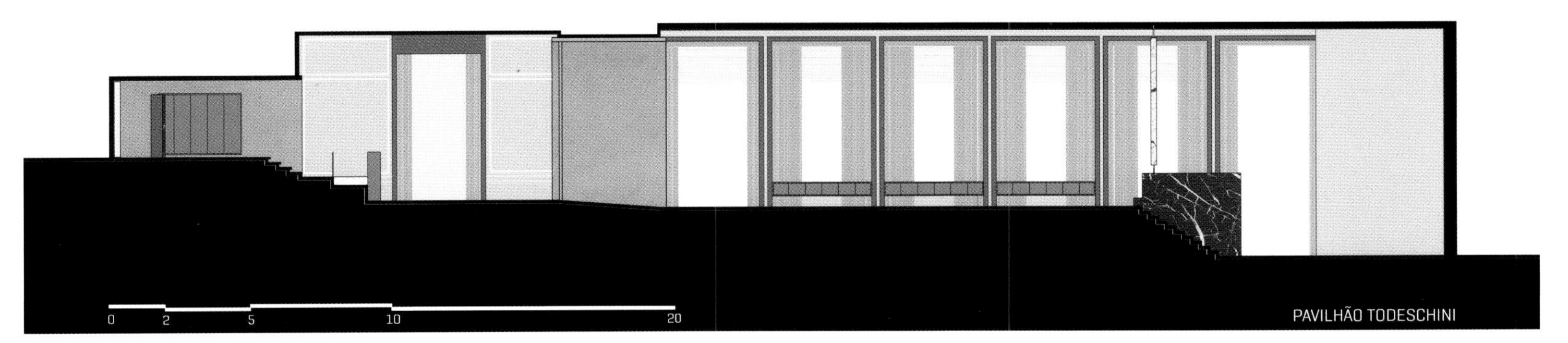
0
2
5
10
20
PAVILHÃO TODESCHINI

Amico

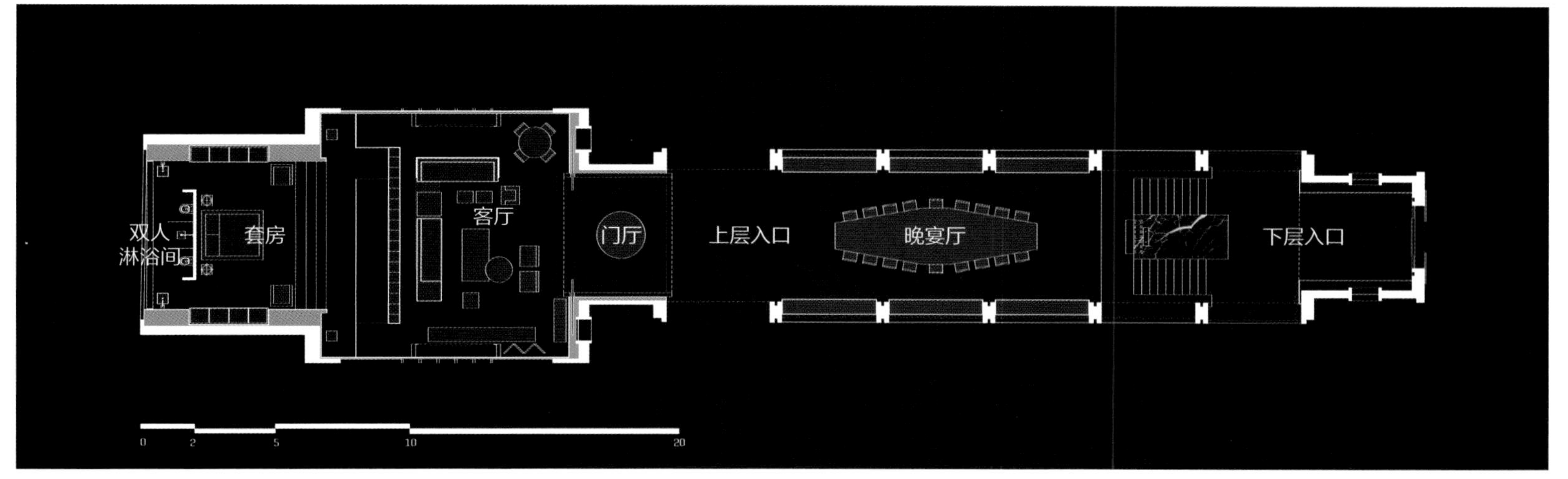
双人
淋浴间
套房
客厅
门厅
上层入口
晚宴厅
下层入口
0
2
5
10
20

The designer creates a simple home space. In order to maintain the integrity of the space, the sliding door is also made in black and white with the book pattern, which echoes with the library behind the sliding door in the model well. At the same time, here is a great end scenery, where is enough to make the owner who is fond of fashion and simplenessand feels his/her heart beating quickly.

The art in the gallery only can be felt to the sensible people. In the modern vision, the bright gold is added into the classical collocation of black and white. The bright colors are shining in the light and interpret the elegance, beauty and gorgeousness of the space in the contrast of black and white.

打造简约家居空间，为保持空间整体性，推拉门同样以黑白色调图书纹路打造，造型上很好地呼应了拉门之后的藏书室，同时此处亦是很好的端景，足以令喜爱时尚简约的你加速心跳。

廊道上的艺术，只有性情中人才能体会。摩登的视觉下，于黑白经典中加入璀璨的金色，明艳的色泽在光线下闪闪发光，在黑与白的半壁对峙中，诠释着空间的奢美华丽。

JEREMY SCOTT
BARCELONA

CHANEL
CELEBRITY BIBLE
BARCELONA

CHANEL
COLLECTIONS AND CREATIONS
ENTREPOSTO
tecidos exclusivos
EDWARD QUINN
minimalismo & color
new minimalist architecture
GIORGIO ARMANI
Tord Boontje
Martina Margetts
MODA É F#%@
VOGUE THE COVERS
BARCELONA

The floor of the whole living room uses the classical collocation of black and white like the black and white keys of the piano to echo with each other and play a flexible variation. In the brightness changes from the brightest to the darkest, it shows a noble and low-key style quietly. The adding of the yellow leather sofa, out of the scale of black and white, enhances the color saturation of the space and presents the elegant and fashionable flavor perfectly. On the one side, the comfortable reading area brings a flavor of literary and art. While the hanging painting with a thinking person on it, is more attractive.

The low cabinet is used as the visual segmentation of the space and is filled with various books which make people feel the flavor of books. At the same time, the sensibility represented by the books breaks through the rationality of the black and white. It is refreshing in the simplicity and introversion. For the people who immerse themselves in the bustling city, a book is the best way to enjoy the fashionable, slow life.

RICK OWENS

整个客厅地面以经典黑白组合，如同钢琴上的黑白琴键，相互衬托弹出灵动的变奏曲。在最亮至最暗的明度变化间，将尊贵低调的型格悄悄上演。黄色皮革沙发的置入，跳脱黑白尺度，提升空间彩度，将优雅与时尚气息尽情展现。一侧舒适的阅读区域，带来了些许文艺气息。而思考中的人物挂画，则更加引人入胜。

以矮柜为空间视线的分割，置满大大小小的图书，书香气息扑面而来。同时以书籍代表的感性，打破黑白的理性，简约内敛中令人耳目一新。对于沉浸在喧嚣环境的人群来说，书籍亦是奏响时尚慢生活最好的方式。

MoMA

Dior
60 Years of Style
MARIA CALLAS
JEREMY

The high-end gray wall allocates with the black carpet, and its junction is separated by the white bedding. The classical black, white and gray create a comfortable and peaceful bedroom like a paradise far away from the bustling city. Its peaceful atmosphere is extremely best for sleep. On the one side, the placement of the bright yellow armchair is exquisite and delicate like rendering the fashionable flavor in simplicity.

高级灰墙面搭配黑色地毯，交界的地方用白色床品加以分隔，经典的黑白灰色调，将卧室空间氛围营造得舒缓而祥和，如同远离闹市的世外桃源，宁静的氛围极其有利于睡眠。而一侧亮黄单椅的放置，于简约中略带时尚，精致细腻。

MUSIC, ARTWORK AND ANTHEM

音乐、画作和赞美诗

摄影师 | ingallery 金啸文
项目面积 | 3000㎡
项目地点 | 北京
设计师 | 闫荣荣、张杰、王元辰、张芝慧、张慧
软装设计公司 | 中合深美
项目名称 | 北京壹号庄园

In the dusk, the majestic voice of hymn sounded, and manor in the moonlight, contains the tranquility of time and the quietness of years. Time at that moment freeze, and a painting of the autumn night is always remembered. When life and architecture blend, life becomes a celebration.

——Balkrishna Doshi

Those People Don't Understand Music Also Shocked by Mozart Symphony No. 41

In 1788, Mozart finished his final masterpiece to his life for the world—Symphony No. 41, and since then, the work has become a miracle in the history of music. It is magnificent and considered to be the greatest orchestral works of humanity; it is outstanding, praised by music critics as God's handwriting, and praised even by those who do not know music at all, while impressed by his magnificent movement. This is a miracle in the history of music, and such a miracle is followed hundreds of years by continuing and developing through different forms of art.

"Music is the flowing construction, construction is the solidification music." Hegel's remark seems to explain the essential connection between these two forms of art in music and construction. Beijing, as the most dynamic city in the world, is most likely to host the brilliance of art. Driving out of the North Fifth Ring, a vast open and lush, fertile land appears there, and it is the Beijing One Manor. Slowly into this piece of land, the bustle in the twenty-first Century seems to fade away gradually. A road leading to the depth of the property seems like masterpieces playing one by one. Through the lush branches and leaves, the villas on both sides appear gracefully. After passing the "moat", at the end of the road, there stands a castle in the forest park and wetland park, which is magnificent. The castle is occupied with a spacious square, covering 3,000 acres, and it is called the manor's "back garden".

In such a noisy era, things have been changing rapidly, but there is a manor house that exists in Beijing with inclusive culture and stands close to the thousand-year-old Imperial City. Here, the conception of aesthetics in life is concealed. It is a piece of land that exists in imagination. Its beauty once energetically existed in the pillar of ancient Rome and Mozart's Symphony No. 41, above the summit of the greatest architectural art in the world. Just as musicians are used to praising Mozart, even those who do not understand the building, with a glance at this magnificent creation, will be shocked.

暮霭中，赞美诗庄严的声音响起，月光中的庄园，包容着时光的静谧和岁月的静好，时间在那一刻定格，一幅秋夜的油画被永远铭记。当生活方式与建筑融为一体，生活便成为一场庆典。

——巴克里希纳·多西

不懂音乐的人也会被莫扎特第四十一号交响曲震撼

1788年，莫扎特交出了人生的最后一部交响曲——《C大调第四十一号交响曲》，自此这部作品成为音乐史上的奇迹。它规模宏大，壮丽灿烂，被认为是人类最伟大的管弦乐作品；它恢弘庄重，气韵万千，被音乐评论家赞叹为上帝的手笔，也被称赞为即使是丝毫不懂音乐的人，也会被其宏广磅礴的乐章所折服。这是音乐史上的奇迹，而这样的奇迹在后续的数百年，通过不同的艺术形式延续并发扬。

“音乐是流动的建筑，建筑是凝固的音乐。”黑格尔似乎诠释了音乐与建筑两种艺术形式之间最本质的联系。北京，作为全球最具活力的城市之一，最有可能承载这种艺术的瑰玮灿烂。驱车驶出北五环，来到一片开阔宏大、水草丰茂的土地上，眼前便是北京壹号庄园。缓缓驶入，21世纪的尘嚣似乎渐去。一条通往深处的主轴乐章般一一奏响，透过繁茂枝叶，两旁别墅群静雅隐现。过了开凿的“护城河”，尽头处有一座城堡，在森林公园、湿地公园中静立，气势恢宏。这座城堡有宽阔的广场，占地3000亩，是庄园的“后花园”。

在喧嚣时代，事物极速变化，却有一方庄园，在北京兼容并蓄的底蕴中，在厚重的千年皇城下存在。在此，生活美学的概念被隐去，它是一片存在于想象中的土地，它的美感曾在古罗马的立柱中，存在于莫扎特的《C大调第四十一号交响曲》当中，存在于这个世界上最伟大的建筑艺术峰顶之上。恰好如同音乐家赞美莫扎特一样，即便是再不懂建筑的人，看到这样的恢弘灿烂，也会为之震撼不已。

The only wide-spread area can be called a manor, Beijing One Manor deserves the name: on a thousand acres of land, 456 villas with an area of about 420 square meters to 2000 square meter hidden in the city's broad spirit. In addition, an area of about five acres to eleven acres of large mansions, their scale is dimensionally wide, making it a unique residence that won't present again, which is solemn and splendid and becomes an honored masterpiece.

宽阔的地域才可被称之为庄园，北京壹号庄园可称为真正的"庄园"：一千亩土地上，面积约420m²至2000m²的456栋别墅安藏着城市的宽广心灵；更有占地约5亩至11亩的11座大宅，尺度之宽广，让其成为当下独一无二的住宅，庄严堂皇，堪称绝响。

Van Gogh Rarely Could Be Understood, But All People Can See The Eternity of Life in *Sunflower*

Van Gogh Museum in Amsterdam, the masterpiece "sunflower" has been quietly awaiting the world to visit. The vitality of Van Gogh's life-long devotion was poured to explain the proposition of "existence" in colourful tunes, and the vitality of the rhythm he devoted to his works culminated in the eternity of his paintings. Van Gogh's paintings were not recognized before his death, but today, more than a hundred years later, we have already assumed the permanent power in the artwork by Van Gogh.

Much like painting, an eternity of architecture also exists in the process of the civilization of mankind, showing the greatest glory. Beijing One Manor is also the remarkable example here. The precious dignity flowing in the buildings is the strong colour from fermented wine and is a dream that relaxes on light and shadow. In the manor, visitors are able to find a story that haunts the space and a daily life afternoon, like the West Grand View Garden, showing its retro and elegant dreams.

Through the analysis of economic society and cultural fields, Rong Chuang gave birth to a new residential culture in Beijing. In the impetuous time, the manor is a bounce-back echo of the "noble" spirit of the new era. The interior emphasizes a sense of spaciousness, simultaneously, the partition and outline of the walls appear elegance in terms of refinement.

Gold is fashionably popular in Renaissance as the noble color. In Beijing One manor, ubiquitous metal lines constantly change, inlaid with gorgeous elegance; paintings on the wall reproduce the glory of the Renaissance era. At this time, people start to think of praised poetry—the rigor and rhythm of the architectural form like the music composed by the exquisite five-line chord, playing the violin music in the morning wake-up.

The bright light and transparent space make people feel the closest distance from the sky, as the Bacchus hanging on the wall, which means that life with wine and rose is daily life. Classic is able to cross the time and space, forming a consistent goodness.

Poetry of Shakespeare Wrote Romance, and Love of Time As Well

From the exquisite palace-like living room into the balcony, leaning on the Roman column, there comes out a gorgeous scene of a movie.

At noon, the French manor dating back to two centuries ago slowly revived, and only existed in the romance of poetic letters and reborn in this piece of land.

In Shakespeare's poems, between the vibrant long and short sentences, this kind of unspeakable romance sounds right, just as the rhythm of Beijing One Manor. In the abundance of time, there lays youth and ideals, freedom and love, pure romance and broad love. They are quiet, but vigorous rush.

Proximately, there is 320-square-meter courtyard lawn in the mansion, even it could be used for equestrian performances, walking in it as much as rambling in the clouds. This elegant activity can only be practiced with enough dimensions in the backyard of this mansion. The character of gentleman rider is not maintained the rigid manners and images in front of people, but rather inadvertently swaying grace and content.

Beijing One Manor, retaining the initial dream, was born with a romantic atmosphere. The sound of cello sounds in the morning mist and the French expression flows through the sound of violin in the sunset. Goblets, small folding fans and all kinds of evening dresses highlight their texture complemented by the white candlelight. Aroma of sweet wine haunts those who stand in the room.

Love is given a test to time, so is life. The manor has described the scenes of life. There, colour is heavily embellished, as well as ambience and light are brought on the horizon. The rhythms it presents, after the contrasting and comparison of the bright yet pure color blocks, form a unique decorative effect and inner strength: the manor contains everything, allowing you to feel the tranquility of time and the quietness of years in every night, just like the power of music, or the brilliance of painting, or the infinite romance in a hymn.

很少人真正懂梵·高，但人皆懂《向日葵》中的生命永恒

在阿姆斯特丹的梵·高博物馆里，《向日葵》安静地等待着来自全世界的游人观赏。梵·高穷尽一生精力，用充满阳光的色彩诠释着关于“存在”的命题，倾注给作品律动的生命力，最终成就了画作的永恒。一百多年后的今天，不同于梵·高生前不被认可，我们早已对梵·高以及他的《向日葵》恒久的力量赞赏不已。

和油画一样，建筑的永恒同样存在于人类的文明进程之中，向所有人展现着最伟大的光彩，北京壹号院亦是此间表率。流淌于建筑之间那份可贵的厚重，是历经时间发酵的红酒般剔透的浓重色彩，是轻盈在光影晕染之间的悠长。庄园里，有回声渺渺的故事，也有午后阳光的日常，如同西方的大观园，展现着复古而又优雅的美梦。

透过对经济社会和文化领域的剖析，融创于北京孕育出了新的居住文化，在浮躁的时代，庄园是对新时代“贵族”精神的回响。室内强调明朗的空间感，墙体分割和轮廓根据精至毫厘的运算，打造出优美典雅的空间感受。

文艺复兴时代流行金色为高贵之色。在北京壹号庄园中，无处不在的金属线条不断变化，镶嵌着华美的高雅；墙面上的画作，重现文艺复兴时代的荣光。此时此处，令人不禁联想到赞美的诗歌——建筑的古典形制严谨而有节奏，就像组成音乐的精妙五线和弦，奏响唤醒晨间日出的提琴乐曲。

明亮的采光，通透的空间，让人感受到与天空最近的距离，如同墙面上的酒神，意味着在此，有美酒和玫瑰的日子即是日常。经典，能够穿越时间与空间，形成一致的美好。

莎士比亚的诗行里书写了浪漫，也书写了时间的爱意

从精美如行宫般的会客厅步入阳台，倚罗马柱，眼前出现电影中的华美场景。

正午时分，两个世纪以前的法国庄园慢慢复活，只存在于诗行之间的浪漫，在这片土地上仿若重生。

莎士比亚的诗篇里，跳动的长短句之间，那种难以言喻的浪漫娓娓而来，这恰如北京壹号庄园的韵律，丰盈的时光所包裹着的，有青春与理想，有自由与爱情，有十足的浪漫和博大的爱意，它们安安静静，却又雄浑奔涌。

大宅中约320m²的庭院大草坪，甚至可以进行马术表演，行走其间如在云中漫步，这项优雅的活动只有在庄园大宅的后院里才有足够的尺度践行。骑马的绅士品格，不是在人前刻板维持的礼仪和形象，而在于不经意间挥洒的风度与内涵。

今日的北京壹号庄园，留存初始的梦想，天生具有浪漫的气息。晨间雾霭中响起大提琴声，黄昏夕照中，法国风情在柔和的小提琴声中流淌，高脚杯、小折扇和各式的晚礼服在白色的烛盏旁突显质感，甜葡萄酒香气袭人。

爱交给时间来检验，生活也要交给时间来检验。庄园描述的是生活场景，渲染的是色彩，表现的是气氛和光线。它所呈现出的韵律是鲜明而纯净的，色块经对比与衬托后形成一种独特的装饰效果和内在力量——包容一切，使你在每一个夜晚都能够感受到时光的静谧和岁月的静好，就像音乐的力量，像画作的灿烂，像一首赞美诗里无穷的浪漫。

FALLING IN LOVE WITH ART AND THE CHARMING HOME

情迷艺术，恋上魅力家园

摄 影 师 | 林峰
项目面积 | 800 ㎡
项目地点 | 浙江绍兴
协作设计 | 徐樑、金俊杰、林娥芳、周煜、王嘉龙
装置与陈列 | 1637
主案设计 | 吴振宝
设计公司 | 安生设计
项目名称 | 绿城玉园私人别墅

"I love to mix styles from different parts of the world. You have to be inspired by your life and that can come from your trips around the world. There is always something from a trip that makes you happy, and you have to use that. Your house is a mirror of your life. You have to take all your experiences from it and put them in the place where you live."

——Patrizia Moroso

Designers believed that a good design is based on humanistic concern, subtly changes our daily life and lifestyles, pursues the transcendental interest and pleases our senses and mood. Through aesthetic transcendence, a good design can save the hearts which are entangled in the center of worldly interest relationships to enhance our taste. At many times, what we need to guide is a kind of lifestyle and attitude, not restrained to a certain style. This case reflects this concept to us perfectly.

“我爱来自世界各地不同地区的风格的混合。你总会从你的生活中得到灵感，比如来源于你在世界各地的旅行。旅途中总有些东西使你快乐，你必须拥有。而你的房子就是你生活的一面镜子，你要把你生活中所有的经历都收藏在你住的地方。”

——帕特里齐亚·莫罗索

设计师认为，良善的设计以人性关怀为基础，潜移默化地改变着我们的日常生活与生活方式，追寻超然意趣，畅快淋漓地愉悦我们的感官，也愉悦我们的心情；通过审美的超越性，救渡纠缠于世俗利害关系中的心灵，提升我们的品位。很多时候，我们需要引导的是一种生活方式和态度，而不是局限于某种风格。本案恰到好处、淋漓尽致地将这一理念体现在我们面前。

THE HOTEL BOOK
KELLY HOPPEN STYLE

BEDMAR & SHI
HOUSE OF HOPPEN

This villa is located in Shaoxing, an ancient city in the south of the Yangtze River known as "the Venice of the East". This building has four floors. The living room, dining room, kitchen and other public activity areas are on the first floor. The drawing room, entertainment area and working area are in the basement. There are three bedrooms on the second floor. And on the third floor, there is the master suite, a cloakroom and the island platform.

别墅坐落在素有“东方威尼斯”之称的江南古城绍兴。四层的空间中，客厅、餐厅、厨房等公共活动区设于一层，会客区、娱乐区、工作区设在地下层，二层设三间卧室，三层则为主卧套房、衣帽间，附带中岛台。

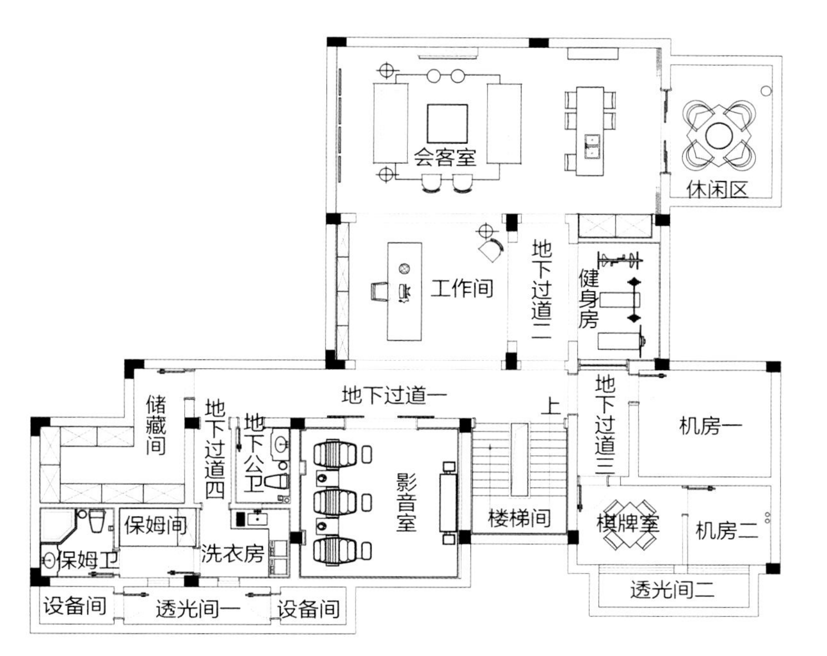

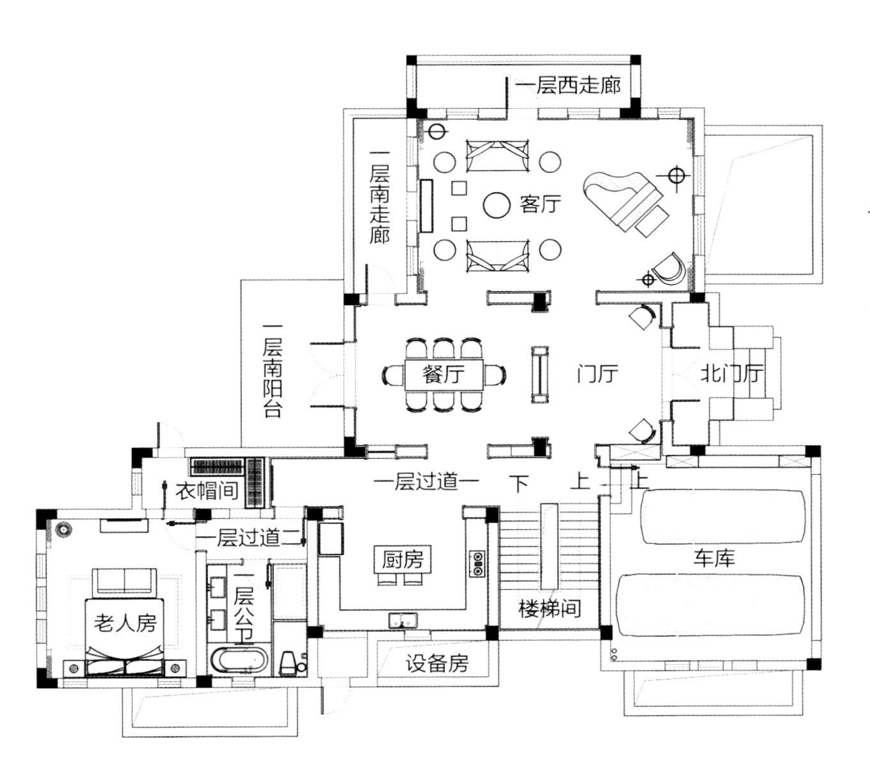

This living space covering 800 square meters has the comprehensive functions. In order to show a more youthful configuration and make the owner's lifestyles and forms tend to be more modern, designers pay more attention on permeability, nature, comfort and convenience in the sense of space. Therefore, designers use the symmetrical door as the boundary of space segmentation to connect the dining room and kitchen on the first floor. The entertainment area on the basement is positioned as a light industrial style, and the open study fuses with drawing room, bar area and fitness area as a whole, which can form a more convivial entertainment area. While the relatively independent but associated video room provides a zone of privacy and ensures the quality of a good audio experience.

空间功能设置齐备，为了呈现更为年轻化的配置，使业主的生活方式和形态也都更趋向于现代感，设计师在空间感受上格外注重通透、自然、舒适、方便。为此，设计师将一层的客餐厅、西厨房打通，仅通过对称的门洞作为空间分割的界定；将地下室活动空间定位为轻工业风格，开放式书房与会客区、吧台、健身区融为一体，形成一个更欢乐的娱乐空间，而相对独立又有关联的影音室又增加一定的私密性，保证享受影音过程时的视听品质。

ROUTE
66
Cine
noi?

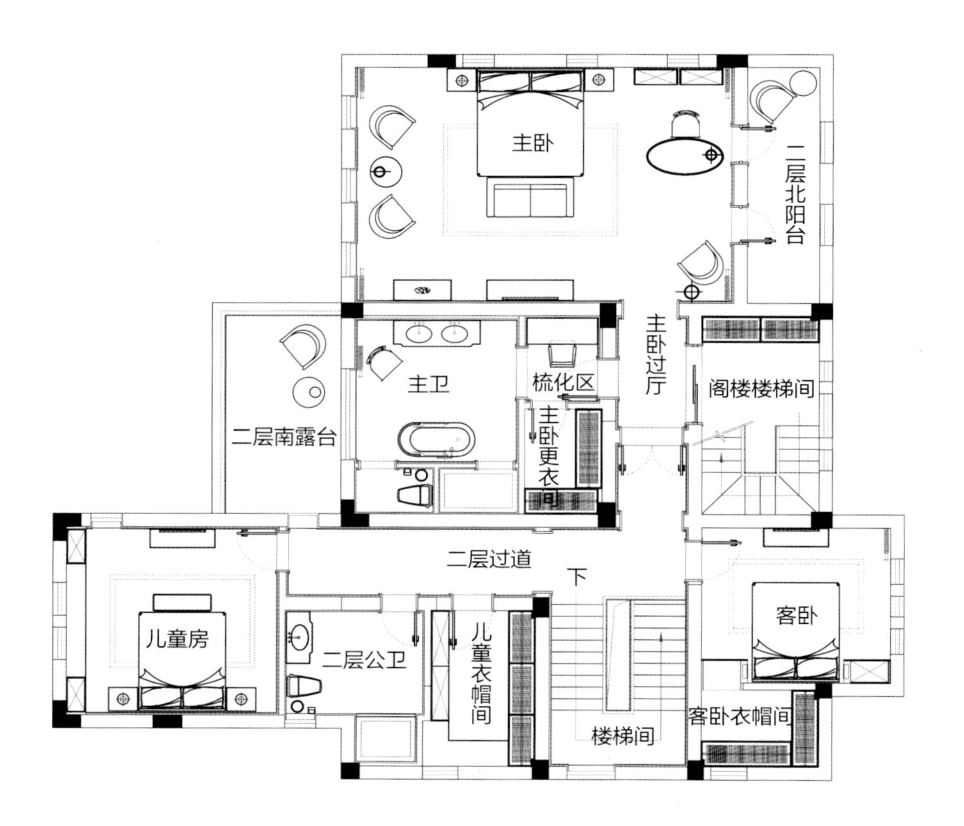

Besides, designers integrate many works from Yayoi Kusama, a Japanese contemporary artist, collected by the owner into space. Anyone who sees these works will have a different feeling. While the endless dots and stripes, and the colorful flowers overlap into the ocean, confusing the existence of a real space. And the dizziness and confusion of where am I are intoxicating.

此外，设计师在空间内融入多件业主珍藏的日本当代艺术家草间弥生（Yayoi Kusama）的作品，令任何看到这些作品的人都会得到截然不同的感受。而无穷无尽的圆点和条纹，艳丽的花朵重叠成海洋，混淆了真实空间的存在，只有阵阵眩晕和不知身处何处的迷惑，令人沉醉不已。

Trainspotting
BEAUTIFUL YOUTH

BEAUTIFUL YOUTH
CENTURY
BEAUTIFUL YOUTH
CENTURY
START SMALL

RENZO PIANO
RENZO PIANO
BEDMaR & SHi

Walking into this single villa reformed from an old house, it seems to be in an amusement park. In the process of design, the designer is in charge of not only the internal design but also the adjustment of the interior structure as well as the planning and transformation of the external building. Finally, the modern yet classic design from inside to outside perfectly presents a new house. In the colorful art space, colors interweave with shapes, and the unique scene elements are fused into a family feast full of interest, which brings the endless reverie.

As soon as you enter the door, you will be attracted by the light on the top. The designer selects the lamp interwoven by blue metal lines with strong plasticity. The elegant lines and the handicrafts full of geometry on the Rome Console Table form an ornament of softness and toughness, which is ingenious and beautiful. While the blue and green carpet with the geometric layout and the decorative painting on the table become a highlight. These furniture ornaments created by the designer have different visual breakthroughs and creativity which can make the whole space infiltrate the representative color extension of the designer. People can shuttle between abstract and concrete space to feel the designer's concentration.

BREAKING THROUGH THE BOUNDS, CREATING A VISUAL FEAST OF A HOUSE

突破界限，经营家的视觉盛宴

摄影师一刘鹰
项目面积一300㎡
项目地点一浙江宁波
设计师一葛晓彪
设计公司一金元门设计
项目名称一丹桂花园

FRANK SKINNER
DAN BROWN
CORNWELL
GEORGE MICHAEL
MUTINY
TONY BENN
chris evans
LANCE ARMSTRONG
DAN BROWN
CORNWELL

Entering into the living room, you will be attracted by the layout of the soft decoration. The designer uses the classical decoration to allocate with the modern Rochebobois furniture, which is colorful, random, modern and interesting. The UFO-shaped table with the luster of black pearl shell makes people have the impulse to sit down and enjoy it. Undoubtedly, the decorative art is the most important ornament of this house. For example, the geometric screen behind the sofa breaks the monochrome situation of every object and makes the space stereoscopic and jumping; round green acrylic lamp is also like an ornament, hanging on the white wall quietly which becomes a beautiful scenery; the slanted herringbone tiles on the ground break the excessive square in the space and embody the designer's skills in details.

The hallway leading to the dining room is also wrapped in parapet. The paintings on the wall designed by the designer make the layout of space clearer, and tell the story what the designer wants to tell. In the dining room, the combination of parapet and paint is harmonious with the atmosphere and makes the proportion of the space more elongated; the orange paint becomes softer along with the refraction of light and shadow; the scattered chandeliers make the space more dynamic. Besides, the designer selects the long dining table with the butterfly tenon which can inadvertently make the owner recall the spirit of elaboration and the attentive mind at that moment in the old time.

Grisham FORD COUNTY

卫生间
老人房
厨房
休闲区
钢琴房
走道
卫生间
客厅
玄关
餐厅

The kitchen is the most nuanced place. The designer thinks that kitchen is also a kind of living culture. Imagining that a family sit around the bar, and the hostess is doing breakfast and chatting with her family, how harmonious and warm the scene is. Therefore, the designer abandons the original space in the back of the kitchen to make the kitchen transparent, bright, open and gorgeous. In the cabinet, the handling of the color separation between the above and below makes the levels clearer, and the Indonesian lacquer painting on the wall becomes a highlight of space. All of these also show that the designer always hopes to explore the relationship of integrating with the design, art and fashion, remove the definition among them, extract the essence from the classical design in the past and apply the technology and the lighting handling of the modern artisan to create a unique spatial temperament.

走进这座由老房子改造的单体小别墅，仿佛来到一家游乐场。设计过程中，设计师不仅要负责内在的设计，还要兼具室内结构的调整以及外部建筑的规划和改造。最终，由内而外，现代而又经典的设计完美地呈现为焕然一新的房子。在色彩斑斓的艺术空间中，色彩与形状相互交织，独具匠心的场景元素融合成一场充满趣味风情的家居盛宴，在繁复之余令人产生无限的遐想。

一进门，就会被顶上的灯光所吸引。设计师选用可塑性强的蓝色金属线条交织而成的灯饰，优雅的线条和玄关几上富有几何感的工艺品构成一柔一刚的点缀，巧妙而美好，而几何版面的蓝绿色地毯和台面上的装饰画又仿佛画龙点睛。这些由设计师自行创作的家具饰品，具有不同视觉上的突破与创意，令整个空间都浸润着设计师标志性的色彩延用，让人穿梭于抽象与具象之间，感受设计师的用心。

步入客厅区域，便会被软装布局所吸引。设计师用古典的装修搭配现代的罗奇堡家具，色彩丰富，随意又摩登有趣。形似UFO的茶几似有黑珍珠般的贝壳光泽，让人有坐下细品的冲动。装饰艺术无疑是这个家最重要的装饰品，如沙发背后的几何图屏风打破了每个物件单色的局面，令房间立体跳跃；亚克力绿色圆形灯饰又像一个装饰品，静置在白色护墙上，风光旖旎；地面上斜铺的人字形地砖打破了空间里过度的方方正正，在细节处体现出设计师的功力。

通往餐厅的过道也用护墙包裹，墙上由设计师自行设计的画作让空间的脉络更为清晰，诉说着设计师想讲的故事。餐厅处，护墙和涂料的结合调和了氛围，让空间的比例更为修长。橘黄色涂料随着光影的折射越发柔和，高低错落的吊灯则让空间更加灵动。此外，设计师选用打了蝴蝶榫的长餐桌，不经意间便会让居者回想起年代久远时的那一瞬间的匠心精神，心思细密。

厨房是最细致入微的地方。设计师认为厨房也是一种居家文化，试想一家人围坐在吧台边，女主人一边做早餐，一边和家人畅所欲言，是多么融洽和温馨的场景，因而设计师舍弃原始格局中厨房后的一个空间，让整个厨房更通透敞亮、开阔大气。橱柜处，上下分色的处理让层次更分明，墙面上的印尼漆画把空间推向高潮，也表现出设计师一直希望探索融合设计、艺术和时尚的关系，去除彼此之间的界限，从过去的经典设计中提炼精髓，运用现代匠人的工艺和光线处理，营造出一种独特的空间气质。

Up the stairs, all kinds of textures and lines models in the stairs outline the space more stereoscopic with lights. Up to the second floors firstly, the family lounge jumps into our eyes, and two rows of cabinets full of various books highlight the owner's preference and taste. The gauze curtain wobbles gently in the breeze, which is beautiful and romantic. The top surface adopts a modern figurative shape allocating with the simple and smooth lines without any nodes, which is elegant, delicate and interesting.

拾级而上，楼梯处各式质感与线条造型随光将空间勾勒得更加立体。踏入二楼，首先映入眼帘的是家庭休闲厅，两排放满各式书籍的柜子突出了主人的喜好以及品位。纱幔随清风轻轻摆动，美妙而又浪漫；顶面采用现代感的具象形状搭配没有太多节点的简约平滑线条，雅致而又富有情趣。

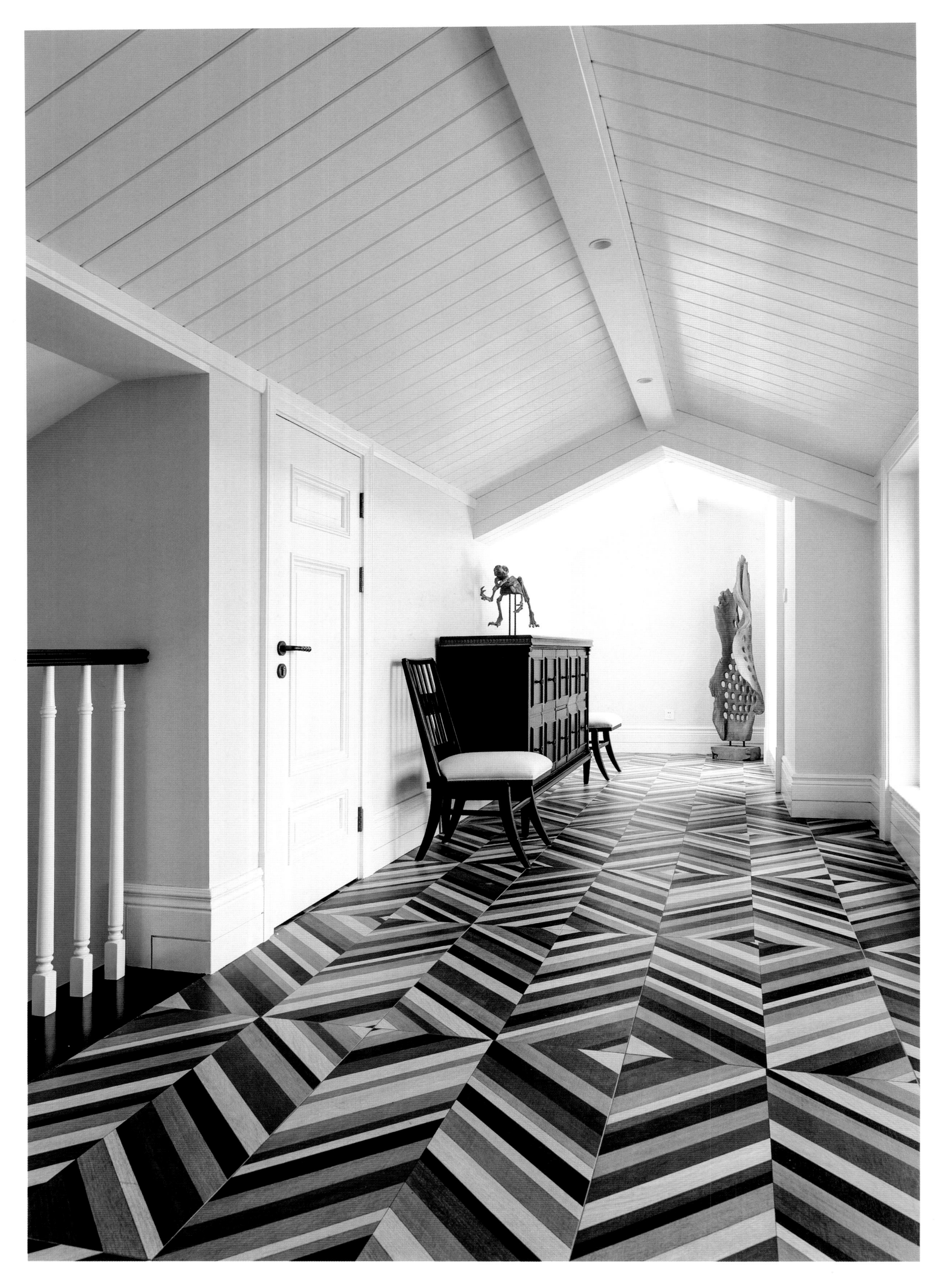

The master bedroom is at the double door on the left of the lounge, the concise lines blur the boundaries, while the classic shadow is seen clearly in details such as the wall, the back of the bed, and so on. The designer uses the color similar with the blue and gray, classic design objects and the contemporary art creations to refurbish the space atmosphere, creates a sense of interest between different elements and uses the contrast and the same color tone of a group of decorations to show a unique living space. On the other side, there are two bedrooms for their sons. The little son's room is lively, and the yellow cabinet with a lattice pattern and the animal-shaped toys are harmonious, making people indulge in the beauty of their childhood. The elder son's room is relatively calm, using the current popular dusty pink to allocate with the British grid bed. Under the reflection of the light, it can make people not only taste its details and textures but also enjoy the unique temperature of this house.

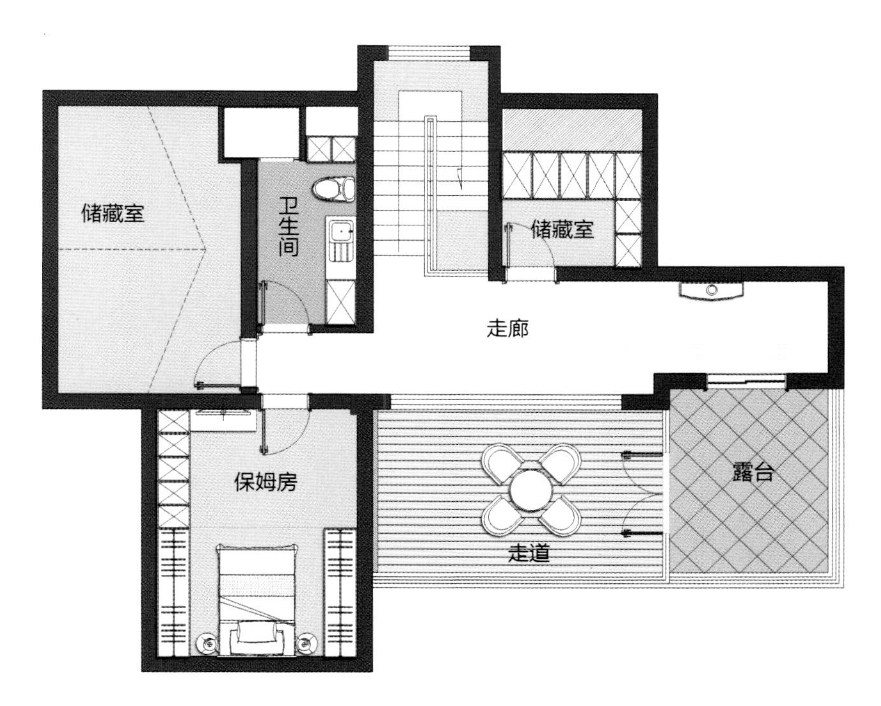

休闲厅左边的双开门处便是主卧，简洁的线条模糊界限，而经典的影子在墙面、床背等细节中清晰可见。设计师启用蓝灰色，运用经典设计单品与当代艺术创作，翻新空间气氛，在不同的元素间创造出趣味感，利用对比性以及同色调的组搭装点出别具一格的居室空间。另一边是两个儿子的卧室，小儿子房活泼俏皮，黄色的菱格斗柜和动物造型的玩具相互协调，让人沉醉在童年的美好中；大儿子房相对稳重一些，运用当下流行的藕粉色，搭配英伦格子的床，在光的照射下，让人反复回味其中的细节和质感，也让人不断回味这个家的独特温度。

TRAVELLING THROUGH TIME TO ENJOY THE UNIQUE SCENERY

步步成景的时空穿越旅途

项目面积｜156 ㎡
项目地点｜北京
设 计 师｜程晖
设计公司｜唯木空间设计

This is an ordinary loft house hidden in the bustling city. The designer demolishes the floor stab the second floor to let a vitreous bridge across space, creating a magical sense of passing through time and space. All superfluous walls are dismantled, and the bedrooms and the bathroom adopt the open design. Therefore, the lighting and sense of space are improved relatively. The whole space takes the quiet black as the main keynote and bright yellow as embellishments to bring the strong contrast impact. While the blue, green and red with high saturation happen to coincide with the colors which David Hockney is good at, who is the owners' favorite British artist.

The designer mixes numerous artworks, Chinese furniture, antique brass bed, old Nordic cabinet, refurbished teak floor and modern design articles, allocating with the modern, retro and up-to-date elements perfectly to create various kinds of emotions and an interesting and unique space.

The pursuit of art in space is endless. The large painting Lie Ning by artist Qin Lingsen, the sculpture Suite Horse by Diao Wei, the oil painting On the Way by artist Liu Bing, the etching Trois Vieux Copains en Visite by Picasso, the work Opening Eyes and Closing Eyes by the young artist Wu Jian'an, the antique mirror, the long narrow table from the Qing Dynasty, the mottled green wooden door from the Republic of China and other elements are presented harmoniously by the designer, describe the spirit and soul of the space, highlight the culture of residence and have the presentable texture. Time seems to condense here, and people in the bustling city can go back to the past quiet time and enjoy the unique scenery.

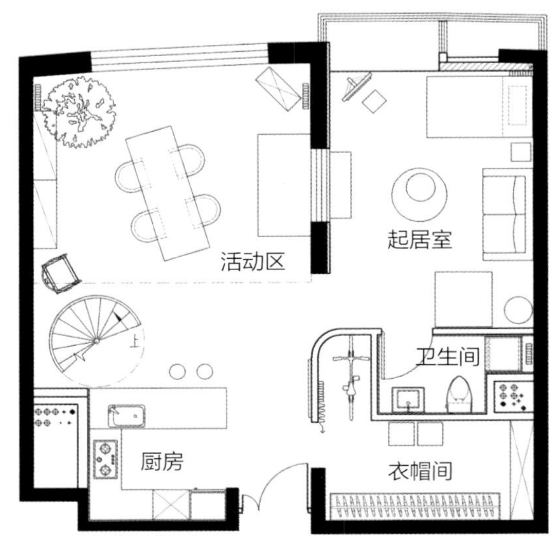
活动区
起居室
上
卫生间
厨房
衣帽间

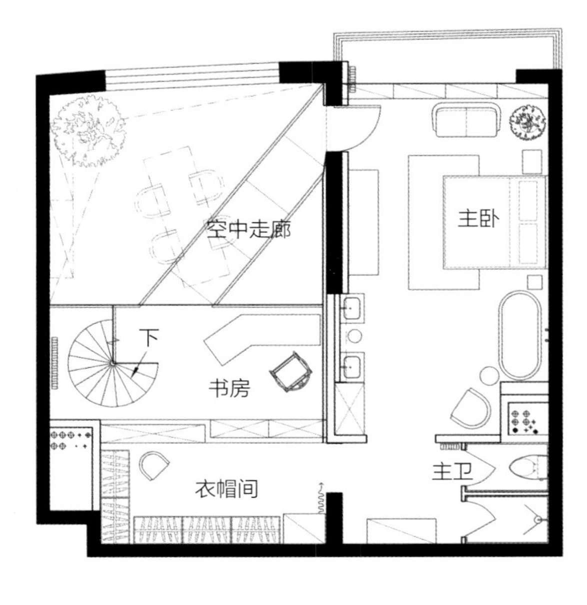
空中走廊
主卧
下
书房
衣帽间
主卫

本案是一套隐藏于繁华都市中的普通loft住宅，设计师拆掉了二层的楼板，令一架玻璃天桥腾空而过，营造出一种魔幻的穿越时空的感觉；拆除所有多余的墙体，卧室和卫生间也采用开放式设计，使得采光功能和空间体验得到了极大的改善。整个空间以静谧的黑色为主基调，以明黄色为点缀，带来很强的视觉冲击，而饱和度极高的蓝色、绿色、红色恰恰与主人夫妇最喜欢的英国艺术家大卫·霍克尼（David Hockney）擅长的色彩不谋而合。

设计师把诸多艺术品、中式家具、黄铜古董床、北欧老柜子、翻新的老柚木地板、现代的设计单品进行了混搭，现代、复古、摩登等元素恰到好处地搭配在一起，创造出许多不同的情绪，造就一个有趣且独特的空间。

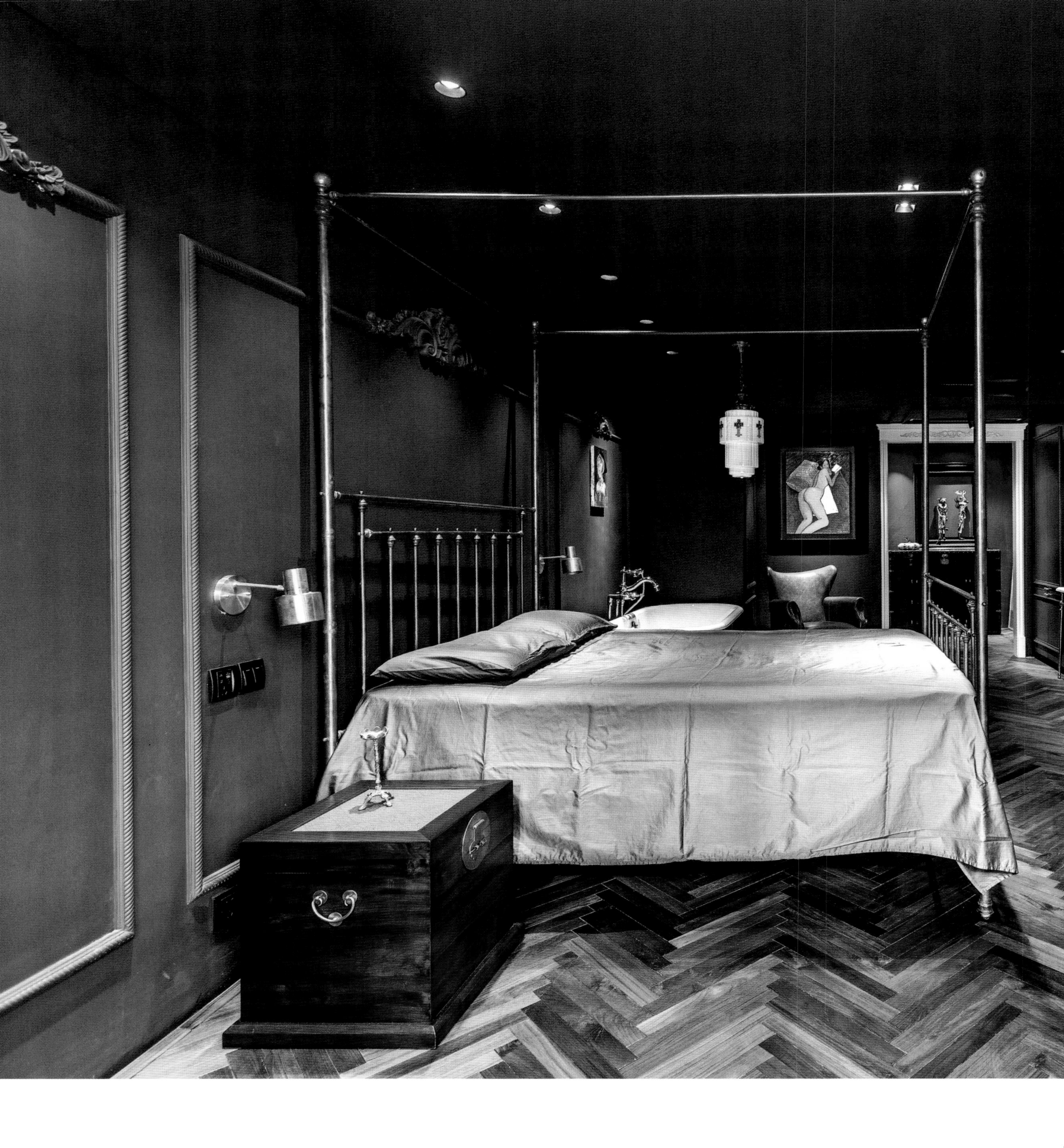

空间中对艺术的追求是无止尽的，艺术家秦铃森的巨幅画作《列宁》、刁伟的雕塑作品《西装马》、艺术家刘冰的油画作品《On the Way》、毕加索的蚀版画《Trois Vieux Copains en Visite》、青年艺术家邬建安的作品《睁眼闭眼》、古董镜、清代的条案、斑驳的民国时期洋行的绿色木门等在设计师的呈现下融洽共处，描绘空间的灵与魂，突显出住宅的文化，有着耐看的质感。时间仿佛在此凝结，让繁忙都市中的人们穿越回从前的静谧时光，独享一处处胜景。

SEVEN
SINGING
SISTERS

THE MODERN SPACE IN WHITE AND BLACK

黑白相生的摩登空间

项目面积 | 120 ㎡

项目地点 | 内蒙古呼和浩特

设 计 师 | 杨昕

设计公司 | 北京王凤波设计机构

项目名称 | 新城名苑A3户型

This is a model house in Hohhot, called "Green City". The designer chooses black and white as the main color tone of this three-bedroom space. At the same time, the designer uses the bright yellow as an interspersed color. In the space, these three colors alternately appear in different sizes, materials and shapes. They can match with each other to show different stunning effects in different functional spaces, or unity and harmony, or calmness, or vitality, and present a visual sense of space full of fantasy and drama.

The dining room is adjacent to the open kitchen, and the allocation of black, white and yellow is fully embodied in these two spaces. The yellow tiles in the kitchen echo with the yellow dining chairs in the dining room, which can make these two functional spaces more unified. The attentive design of breakfast bar also forms a change with high and low scattered arrangement to space.

The children's room with small area still continues the design technique of black and white allocating with yellow. The stars wallpaper full of childish and the paintings in all sizes reveal the functions of the room, while the novel design of bedside niches adds a vertical depth feeling and interest of this small space.

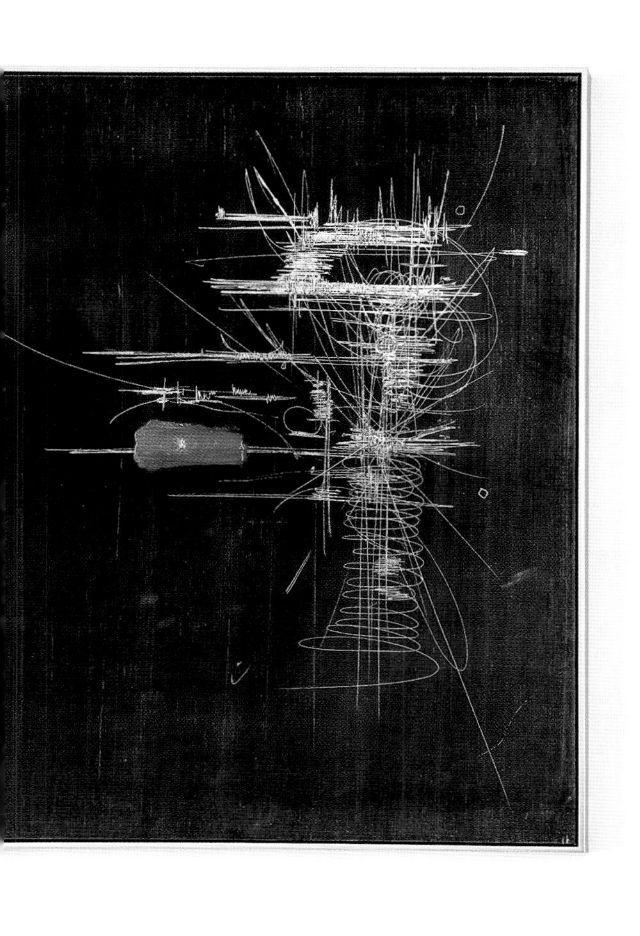

Stylehome
New looks
Brightest
Brightest

这是一套位于“青城”呼和浩特的样板间，设计师以黑、白两色作为三居室空间的主色调，同时以明艳的黄色为点缀色。三种色调在空间中以不同的大小、材质和形状交替出现，在不同的功能性空间中搭配出不同的惊艳效果，或统一和谐，或沉稳，或灵动，呈现出充满幻想与戏剧性的空间观感。

餐厅与开放式厨房相邻，黑、白两色与黄色的搭配在这两个空间里体现得淋漓尽致。厨房的黄色墙砖与餐厅的黄色餐椅形成呼应，让两个功能性空间更加统一。贴心的早餐吧台设计，也给空间形成了高低错落的变化。

在面积不大的儿童房中，黑、白搭配黄色的设计手法依然延续。充满童趣的星星壁纸和大小装饰画，阐明了房间的功用；而新颖的床头壁龛设计，增加了小空间的纵深感和趣味性。

In the elegant little study, the designer makes uses of the changes of color blocks on the walls and geometric shapes to keep unity with the overall design, full of youthful vitality, which cater to the young customers' needs of functions and aesthetics.

在小巧别致的书房中，设计师利用墙面色块与几何形体的变化，使其与整体设计保持统一，在表现年轻活力的同时，又贴近和满足年轻客户群的功能以及审美需求。

Knowledge makes humble,
ignorance makes proud.
Elegance
Brightest
New looks

In the master bedroom, the designer takes the white color as the main color, and the gray as the cushioning color, while the gold is convergent to become the decorative lines on the furniture. Not only that, the extremely personalized furniture and decorations also become the result of the designer's unified consideration and meticulous collocation.

In the bathroom, the allocation of black and white makes the space more spacious and cleaner. The mirror with a unique shape and the yellow tiles in the shower room continue the striking contrast and the proportion of bold design of this whole space.

在主卧室中，设计师以白色为主色调，以灰色为缓冲色调，而金色收敛成为家具上的装饰线条。不仅如此，极富个性的家具和装饰品，也都是设计师统一考虑、精心搭配的结果，塑造出无与伦比的空间质感。

卫生间里的黑白搭配令空间显得格外宽敞而干净，造型别致的镜子，淋浴房中黄色的墙砖，无一不延续着整个空间的鲜明对比与大胆比例设计，突出空间的前卫感。

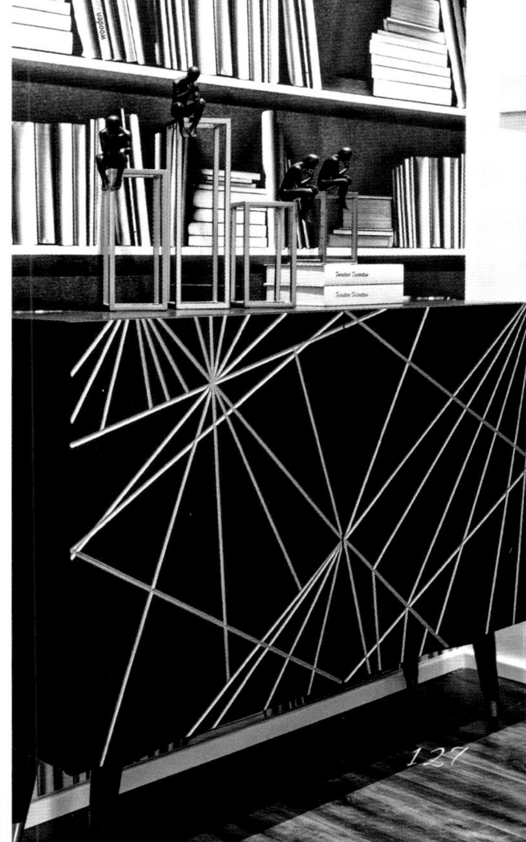

FAITH
mysterious
FREE
HEY
YOU

SHAPES AND COLORS, THE ETERNAL LUXURIOUS AND ELEGANT ART

项目面积 | 500 ㎡
项目地点 — 广东深圳
设 计 师 — 陈列宝
设计公司 — 进革室内 深圳市陈列宝室内建筑师有限公司
项目名称 — 深圳华侨城波托菲诺纯水岸别墅

形与色，永恒的奢雅艺术

This case is an old house in the Riverview Villa reformed by the Studio Revolution, located in the Portofino Lake with the excellent location and scenery. The owner has a high-grade taste and an excellent international aesthetic perspective. After several communications with the owner, the designer positions the design style of this house as the contemporary artistic style.

The entrance hall combines functions with aesthetics through a group of highly artistic collocations. The white European pattern in the living room serves as the cultural texture, with the excellent colors and the great sense of times to show a contemporary artistic interior atmosphere. Most of the soft decorations in this project are purchased from Britain, Denmark and other countries. The theme wall with a very strong sense of sequence in the dining room is installed by the high-end custom team.

Some bedrooms among these three floors adopt a perfect collocation of pure wool and linen texture, allocating with the Turkish handcrafting carpet and the Belgian solid wood floor. The design of bedrooms focuses on the relationship between people and texture as well as the sense of comfort. The public aisle and staircase use a large area white paint and light wood color to create a quiet and peaceful atmosphere. The custom copper drawing stair handrail is an embodiment of the quality and detail.

ARCHITECTURE NOW!
EAT SHOP DRINK

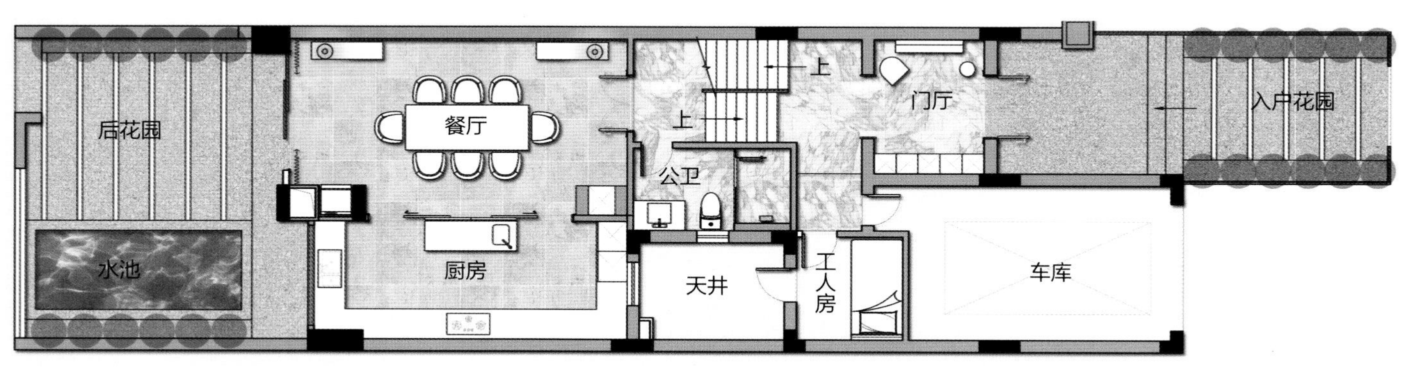
后花园
餐厅
上
上
门厅
入户花园
公卫
水池
厨房
天井
工人房
车库

这是进革室内完成的一套位于纯水岸的旧宅改造作品，位于波托菲诺湖边，地段优越，风景绝佳。业主是品位高端，有极佳的国际化审美视角的客户，在与业主多次沟通后，将本案设计定位为当代艺术风格。

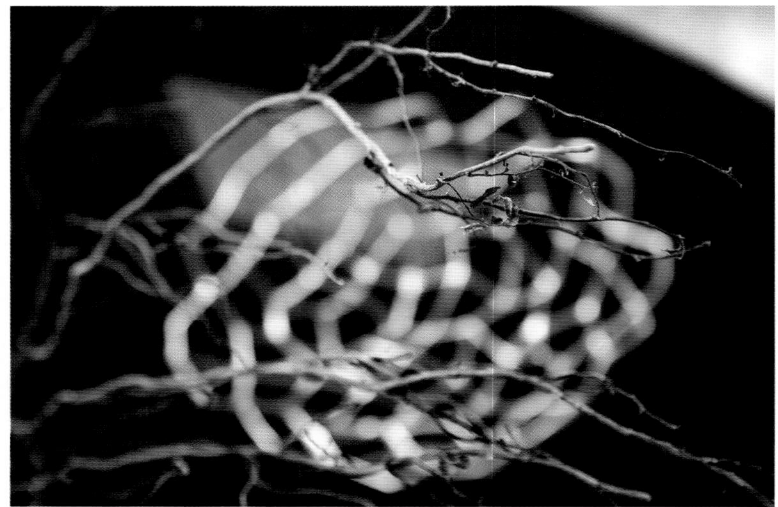

入户门厅通过一组艺术感极强的搭配将功能和美学相结合。客厅内白色的欧式纹样作为文化的肌里，颜色和时代感俱佳的现代家具，共同体现出当代艺术的室内氛围。本项目大部分的软装饰品均从英国、丹麦等国原装采购。餐厅序列感极强的主题墙造型全部由高端定制现场安装。

三层卧房部分采用纯羊毛与亚麻质的绝佳搭配，配合土耳其的手工地毯和比利时的实木地板，关注人与材质的关系和舒适度是卧房设计的重点。公共过道及楼梯部分用大面积的白色涂料和浅色的木色，营造安静祥和的氛围，古铜拉丝订制的楼梯扶手是质感和细节的体现。

The master bedroom chooses the elegant, calm gray and white color tone as the main tone of the space; the champagne gold Southeast Asian handmade decorative boxes and ornaments reflect an exquisite feeling; the blue furniture and carpets make the overall space more active. Opening the door, the independent bathtub is presented at the end of the scenery in the master bathroom as an artwork, and the mosaic walls matching with the beach oil painting create an atmosphere of bathing in the sunshine. The custom installed hand-washing table matches with the creative design of the overall style.

主卧房选择优雅平静的灰白色调作为空间的主调性，香槟色的东南亚手工打造装饰盒与饰品体现精致感，蓝色的家具及地毯让整体空间多一份活跃。开门后，独立浴缸作为艺术品般呈现于主卫生间的端景处，马赛克墙面配合沙滩油画，营造一种沐浴阳光的氛围，订制安装的洗手台由设计师配合整体风格创意设计。

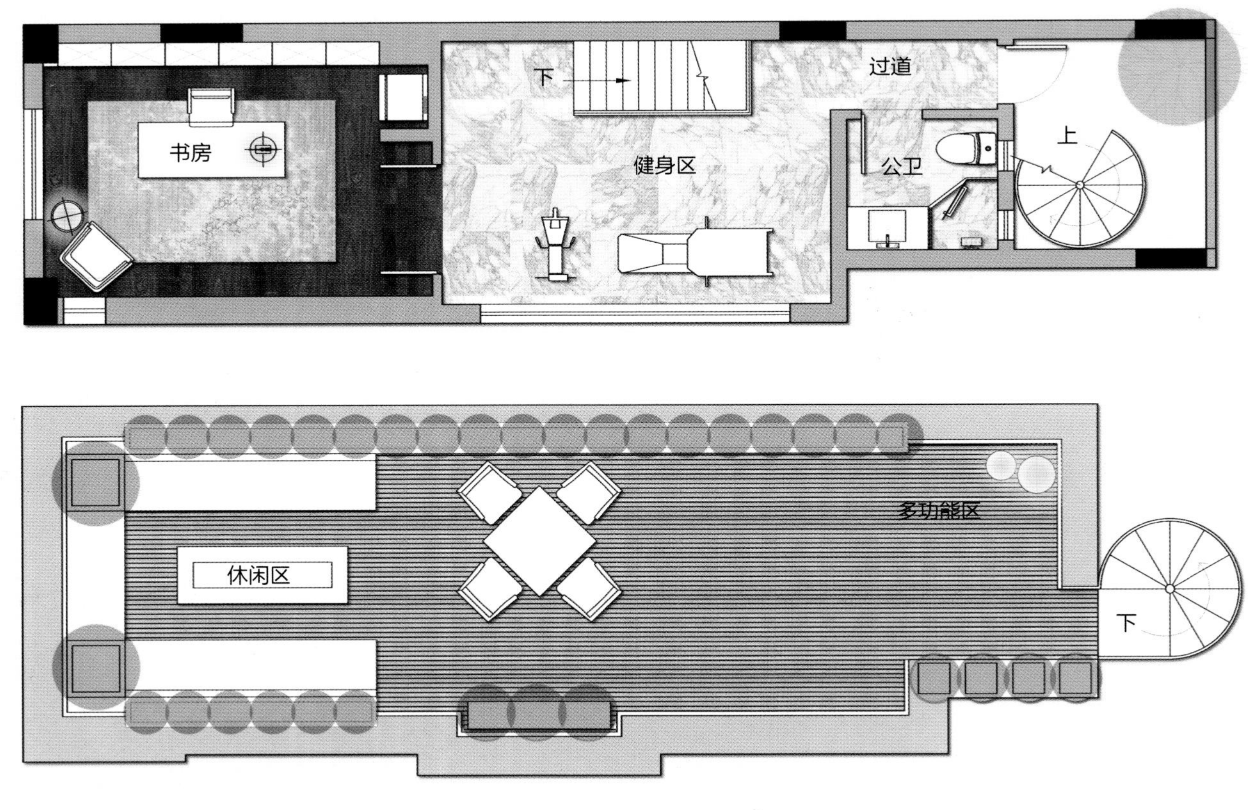
下
过道
书房
健身区
公卫
上
多功能区
休闲区
下

The Vine
Project

THE SPACE FOR CREATING DREAMS

造梦空间

Project Name | Maisonette P155

Design Company | Ippolito Fleitz Group

Designer | Ippolito Fleitz Group

Project Location | Stuttgart, Germany

Area | 290 ㎡

An architect and a textile designer have created a sanctuary in a listed Wilhelminian building in a sought-after location on the edge of Stuttgart's city centre. Their new apartment of 290 m² stretches over two floors with an unusual tapering floor plan that resembles a slice of cake. The apartment has been transformed into a vibrant cabinet of curiosities, filled with mementoes and inspirational pieces, which they have collected or sourced on their travels. A characteristic period feature of the building is its layout of individual rooms grouped around a central hallway. This layout was carefully modified, respecting the building's listed status to create a spacious, open discourse with shifting vistas and overlapping perspectives.

The upbeat is given by a pale gray, gallery-like hallway, which forms a cabinet brimming with travel curiosities. A striking element is a wooden bench from India, which draws you into space, accentuating the suction effect of the trapezoidal layout. A black, herringbone parquet floor runs from here throughout the apartment, giving the suite of rooms a flowing feel and creating a strong graphic counterpart to the typically bourgeois Wilhelminian architecture.

MVRDV Buildings
CONCRETE

A

PROPORTIO

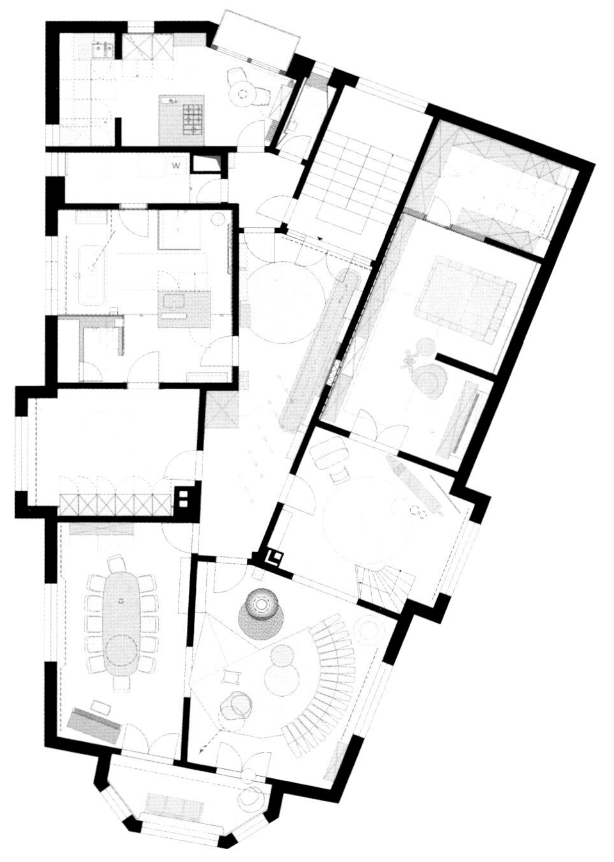

At the head of the hallway is the living room, a salon-like space with strong contrasting colors, intense graphic elements and large forms. A lemon yellow bookcase is positioned against powder blue walls. A deep pile rug with a bold, geometric pattern in strong colors and a Moustache chair are more works of art than pieces of furniture, yet even these are outdone by the expressive pictures and objects on the walls. Two circular and intersecting shapes on the ceiling take over the circles theme, which is echoed at multiple points in the room, as well as spotlighting life below.

The dining room is dominated by textile materials such as a dark green silk wallpaper and others found from the owner's exotic travels, including Uzbek ikat cloth, Indian silk embroideries, Laotian textile applications and African Losa basketwork. A ceiling mural by Alix Waline brings an additional dynamic to space. In the centre of the room stands a large rosewood table, about which various chairs are gathered. One end of the table top is lacquered black. This shiny reflective surface creates a bridge to the piano as well as to a smoked oak sideboard hanging on one wall. Its partially black lacquered front resembles a fragmented mirror, and dissolves the solidity of its form. A hand-crocheted curtain of paper yarn picks up the textiles theme in a more abstract way and provides a fascinating contrast to the elegant, heavy, putty-coloured cotton velour.

在斯图加特市中心边缘的一块寸土寸金之地，一名建筑师和一名染织设计师在一幢受文物保护的古建筑当中，找到了他们的新归宿。这是一套上下两层的天顶阁楼，平面图形似切下的一牙蛋糕，总面积为290m²。整套公寓就像一座有生命的宝藏，充满了主人在旅途中或艺术品市场收集的各种纪念品和灵感源泉。这幢建筑的原有特点是：以中央走廊为轴，多个房间循序排开。鉴于文物保护的原因，这一结构仅略加改动。尽管如此，一个宽敞通透，层次分明的生活空间仍在此诞生。

首先是浅灰色的门厅。这里好似一间画廊，摆放着各种旅途纪念品。一张来自印度的木质长凳构成空间主体，顺着走廊的方向进一步加强了空间原有的锥形产生的拉伸效果。整套公寓内均铺设黑色人字形实木地板，既使房间的结构分割看起来更加流畅，又与建筑原有的保守派的古典复兴主义风格形成图形化的对比。

走廊的尽头是起居室。这是一间近似于沙龙的房间，运用了浓艳的对比色、强烈的图形元素以及大尺度的形态。淡蓝色的墙边站立着柠檬黄色的书架；大尺度几何形状的高绒地毯颜色鲜艳；再加上摆放的Mustache椅，整套组合与其说是家具，不如说更像是艺术品。尽管如此，墙上的艺术装置和画作还是更胜一筹。天花板上两个椭圆形彼此相切，呼应了空间里一再出现的圆形主题，同时也像一盏射灯一样观察着这里的一切。

餐厅大多采用了织物材料，例如深绿色的真丝壁纸，还有乌兹别克斯坦伊卡特布（Ikat）、印度真丝刺绣、老挝补花，以及非洲罗萨（Losa）编织等业主在旅途中收集的织物。统治天花板的是一幅Alix Waline的画作，为空间赋予动感。房间中央放置着一张巨大的蔷薇木桌，四周的餐椅各不相同。桌面的一部分被漆成黑色，其光洁的表面与钢琴和墙边悬挂的烟熏橡木低柜相呼应。局部的黑漆构成镜面效果，打破了家具材质的局限。手工纺织的纸纤维窗帘以抽象的方式再一次呼应了织物这一主题，同时与沉重典雅的调和色系的纯棉平绒形成对比。

The dining room and salon are connected at their far ends by a small room with a bay window. Here the graphic character of salon and the textile theme of the dining room merge in a specially commissioned psychedelic wallpaper, which challenges the eye and forms a provocative backdrop to several colourful artworks. A contrast comes from the more subdued, natural materials world of the furniture and the intense light that is filtered into the room through golden Venetian blinds, which shine brightly in the sun.

An asymmetric, curved wall opening in the opposite end of the salon leads into the staircase room, the only room in which the original oak parquet floor has been preserved. The walls of this room are papered with an English, hand-printed wallpaper featuring an opulent, jungle motif. The exotic atmosphere is heightened by a life-size wooden horse, an archaic artefact from India, which stands before a dark gray smoked glass wall. Creating a first connection to the upper storey, two suspended lamps emerge from a ceiling opening to hang above the horse, almost like a rider. The upper storey is reached via a staircase with indigo treads and a dark green stringer.

St Katharina

SALVE
don't
kill me

From the staircase room, a second double-leaf door leads into the bedroom, which is also a library. A floor to ceiling bookcase covers the longitudinal wall and draws your gaze into the room. A mirrored wall leading to the dressing room underscores this impression of depth. The dark wood of the bookcase and sideboards coupled with the elegant color of the walls give the room a delicate feel. A silken Berber rug and the leather of the bed bolster the quiet and elegant impression of the space. A concealed door in the mirrored wall leads into a dressing room, which contains two large white hanging wardrobes. Two circular and incised areas of glass dispel the volume of the furniture.

To the right of the hallway lies a spacious bathroom. The salmon-coloured design is in harmonious dialogue with the limestone of the floor and several walls. Multiple mirrored surfaces expand the space and create optical bridges to the other rooms by means of reflections. A freestanding washstand made from rosewood with a superimposed mirror unit form a strong centrepiece, of which are grouped a freestanding bathtub and walk-in shower. The black wooden Venetian blinds and a black dotted pattern on the ceiling provide some necessary contrasts in the otherwise soft atmosphere.

The bathroom connects through to a gym, which doubles as a guest bedroom. Lemon yellow walls fade into a white ceiling and suffuse the room with energy. A floor to ceiling closet provides storage and conceals a fold-out guest bed. While its mirrored front is the perfect backdrop to for your daily workout.

Cooking with friends is one of the owners' passions. So the kitchen at the other end of the apartment has stainless steel, industrial-style kitchen block at its centre. Original tiles on the floor and wall provide a scintillating contrast to the precise, sharply edged, solid surface, built-in cupboards. A freestanding marble-topped table offers space for more intimate gatherings.

The guest bathroom is located next to the kitchen. This small room with its many wall-mounted pipes was paneled to create a clean, polygonal shape. The folded effect of the walls is dissolved by a geometric mural. A softly curving mirror provides a welcome contrast and also expands the space.

The upper storey houses a spacious workspace and private TV lounge. The light-flooded top floor also access to a generous terrace with a view of the treetops in the neighbouring avenue. A stunning view over Stuttgart is visible in the other direction. The green theme is programmatic here, which the room is filled with succulents of all shapes and shades of green. A bed nestled in one of the dormer windows offers space for additional visitors. An interior bathroom with a steam shower and generous visibility into the room and to the outside creates a sensual centrepiece.

The maisonette is a museum of memories and a showroom for the creativity of its owners in one. In place of a closed and consistent aesthetic, the apartment functions as a collage of variegated moods. Yet, in spite of their seeming disparity, a synthesis is achieved that perfectly reflects the personality of the owners in the individual rooms.

餐厅和沙龙的尽头由一个悬楼构成的小房间相连，这里将沙龙的平面构成风格和餐厅的织物主题融汇到一起，变幻出一幅业主亲手设计的迷幻壁纸。壁纸的图案既引人入胜，又为空间当中摆设的艺术品提供了非同寻常的背景。穿透金色百叶窗的强烈的阳光让空间焕发生机，再加上家具平静的天然材质，产生出奇妙的对比效果。

沙龙的另一头，一个不对称的拱形门洞将视线引到楼梯间。这是整套公寓里唯一一个保留了原有地板的房间。墙面张贴着手工印刷的英伦风格壁纸，图案是大尺度的热带植物。一匹来自古印度与实物等高的木马站在一面半透明的灰色玻璃背景前，更加强了这个房间的异域风情。木马上方悬垂的吊灯从天花板的开口垂下，好似一名骑手，构成了公寓上下两层之间的一个衔接。通往上一层的楼梯采用深紫色的踏板，搭配墨绿色的梯身。

从楼梯间到卧室由一扇对开的大门隔开。卧室同时也是业主藏书之所。与房间等高的书架悬挂在房间的纵向墙壁上，将视线拉进房间内。房间尽头的衣帽间由镜面隔开，令视觉效果更加深远。深色的实木书架和低柜，以及雅致的墙面色彩令卧室气氛柔和舒适。一块柏柏尔地毯和床身的真皮材质更进一步加强了房间平静优雅的风格。镜面的背后是衣帽间。衣帽间内是两个巨大的白色吊柜。两块相切的椭圆形镜面减轻了家具的视觉重量。

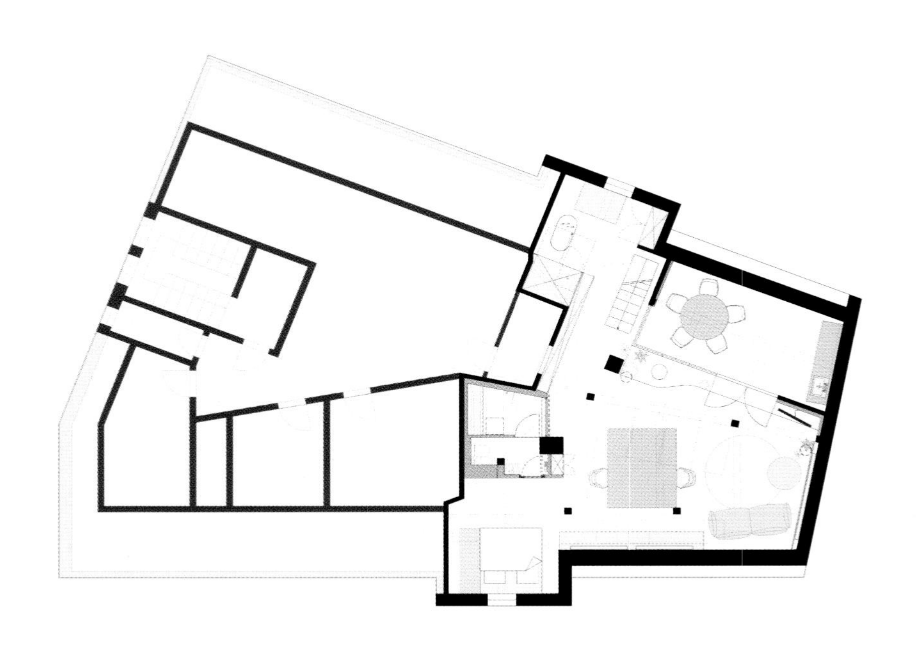

走廊的右手边是一间宽敞的浴室。整个空间采用橘粉色调，与石灰石地面和墙面相得益彰。大面积使用镜面扩大了浴室的视觉空间，映射效果使浴室与其他房间之间产生联系。盥洗台采用蔷薇木打造，隐藏柜脚，与其上的镜面构成空间的中央主体。四周还有一个独立放置的浴缸和一间淋浴室。黑色的木质百叶窗，以及天花板上的黑色点阵，与柔软的环境色之间形成对比。

THE END IS HERE

SQUELETTE DE CHAT

浴室一侧是健身房，健身房同时也可用作客房。柠檬黄色的墙壁渐变过渡到天花板上，为房间注入生命力。与房间等高的壁柜提供储物空间和一张折叠客床。另外，墙上的镜面也为每日的健身运动创造了理想的条件。

业主热爱和朋友一起烹饪。所以，公寓另一头的厨房里有一个不锈钢材质的工业风炉台。厨房地面和墙面的瓷砖保持了建筑的原貌，与简洁直线的合成材质橱柜形成别具一格的对比。一张漂浮结构的圆形大理石桌为少人用餐时提供了适当的场所。

厨房一侧设有一间客卫。为了在有限的空间里遮挡各式管道，墙面采用多边形挡板包裹，几何形态的图案减轻了墙面的折叠效果，曲线外形的镜面中和了空间的硬度，也起到了扩大空间的效果。

楼上还有宽敞的工作间和一个隐秘的休息室。顶楼阳光充沛，连接着一个巨大的露台。从露台望出去，是旁边林荫大道的一株株树冠。工作间的另一侧可以俯瞰斯图加特。绿色是这里的主题，大大小小、形形色色的多肉植物和绿色调充斥整个空间，一张依偎在窗边的床让更多客人可以留宿在此。内置有蒸汽淋浴的浴室视线通透，可一直望到窗外，构成一个散发着感性气息的视觉中心。

这套阁楼公寓是一个承载记忆的博物馆，同时也是展示业主创造力的舞台。这里没有单一的审美，而是一幅融汇了各种风格的拼接作品，虽然画风各异，但却又浑然一体。

A FREE AND AVANT-GARDE RESIDENCE

率性自由的前卫住宅

Project Name I Batman

Design Companies I Ando Studio, Arik Ben Simchon

Designers I Ando Studio, Arik Ben Simchon

Project Location I Itlay

Photographer I Ando Studio

This project is designed in contrast to the old church structure in the Italian architecture look and the modern furniture. All furniture is designed and manufactured by Arik Ben Simchon, a modern eclectic style of design. Because of the structure-dominated, the choices of colors are quite neutral. The jumping yellow and the greenery add vitality to space. The naked stone bricks extend from the corner to the ceiling, and its mottled patterns echo with the retro colors perfectly. The modern batman cartoon doll shows a sense of justice, and makes the whole space brisker. Designers integrate the creativity and ingenuity into space through the stack of simple but natural elements. Its beauty is presented among the original and plain feeling without any pretentiousness, the free personality and concise demands, as well as the fashionable tension and exquisite enjoyment hidden in its rough appearance. Designers integrate the feeling of beauty with the enjoyment of art, then, integrate them with the life of modern people to create a unique aesthetic interest.

THE MOTION MAKER
ARCHITECTURE

ARCHITECTURE

Feel secure in your home
pierre cardin

该项目是通过意大利建筑风格的旧教堂结构与现代家具相融合而设计的。所有家具的设计和制造都来自Arik Ben Simchon——现代折衷主义的设计风格。以结构为主，颜色的选择相当中性。偶尔跳跃的黄色与绿植，增加了空间的活跃度。裸露的石砖从墙角一直延伸至天花顶，斑驳的纹路与复古的色泽形成完美契合。极具现代感的蝙蝠侠卡通公仔的加入，正义感十足，让整个空间多了份灵动。简单而自然的元素堆砌，将创意与巧思融入其中，它的美感，在于不多矫饰的原始和质朴感受，在于极致展现率性与自由的个性特质与简洁诉求，也在于粗犷外表下掩藏其中的时尚张力与精致享受。设计师将美的感受与艺术的享受融合，而后融入现代人的生活，创建出独特的审美意趣。

THE ROMANCE AND LIGHT LUXURY THROUGHOUT ANCIENT AND MODERN TIMES

横亘古今的浪漫简奢

Project Name I Residence in the Italian Countryside

Design Company I Diff.Studio

Designers I Vitaliy Yurov, Iryna Dzhemesiuk

Project Location I Italy

This is a countryside residence in Italy. The interior design is an exquisite combination of classic and contemporary style.

The living room is dominated by white color. Using the dark green suedes sofa in this white environment, this color collocation creates a peaceful and forest-like transparent feeling. While the large fabric chandelier gives people a dynamic sense of blooming flowers. The combination of stillness and movement strengthens the generous, fashionable temperament of the space.

The lines of the dining room are neat and clear, and two ring metal chandeliers in the vertical environment become the visual focus. Space uses a large area crimson as the base, and the classical color collocation of red and black is very fashionable and eye-catching, which is a perfect combination of aesthetics and functionality. While the kitchen with a large pattern is concise and generous, reflecting the owner is an exquisite person who is good at cooking.

FOR THOSE WHO COME FROM NOTHING
TOP 100 INTERIORS

本案例是意大利一套乡村住宅，室内设计是古典与现代风格的完美结合。

客厅整体空间以白色为主调。在白色的大环境中使用墨绿色仿鹿皮绒面沙发，色彩搭配上营造出宁静通透如森林般的感觉，而大型布艺吊灯则给人以鲜花绽放的动感，动静结合中增强了空间知性大方的时尚气质。

餐厅线条整洁明朗，纵线环境中的两盏环形金属吊灯成为视觉焦点。空间采用大面积深红色为基底，红黑搭配的经典撞色非常时尚夺目，完美地结合了审美要求和功能性。而大格局的厨房则简洁大方，可以看得出来居者有一双精致能干的厨房巧手。

The design of the bathroom has to be mentioned here. The overall design is dominated by the collocation of black and white. The walls shaped by the new materials are outstanding, and the clean and bright space gives people an advanced experience, like time travelling.

The artistic intarsia design of the parquet, golden pieces of furniture, as well as the combination of the vaulted ceiling and the beautiful old architecture create an unimaginable and unexpected visual enjoyment. The integration of classic and modernness also creates a unique and unforgettable atmosphere.

不得不提的是本案中的浴室空间设计，整体设计以黑白搭配为主，新材料塑造而成的墙面光彩夺目，光洁明亮的空间给人一种时光穿梭般的超前体验。

拼花的艺术图案设计、家具的金片、拱形天花板与漂亮的老建筑结合，迸发出不可思议且意想不到的视觉享受。古典与现代相结合，二者共同营造一种独一无二的、难以忘怀的气氛。

A GENTLEMAN SPACE FULL OF ELEGANCE

富于雅趣的绅士空间

Project Name I Colored Pattern Story

Design Company I Diff.Studio

Designers I Vitaliy Yurov, Iryna Dzhemesiuk

Project Location I Kiev, Ukraine

In this apartment, the accents of interior design are the various patterns and ornaments as decorations. Designers use prints and repeating patterns in textiles, carpets and other accessories to achieve an unusual visual effect and emphasize a personalized atmosphere. At the same time, except using the unusual patterns, designers are also good at using the color collocation, create an unconventional space through the contrast design and use the unique furniture which can fully present the owner's characteristics and preferences to create a harmonious and prominent indoor environment.

The variety of natural materials and the abundance of greenery seen everywhere in the public areas give a sense of comfort and show energy and vitality in the space of deep gentlemanly demeanor. The pig-shaped salver in the bedroom adds a touch of bright jumping color, and the creative allocation of black and white enhances the texture of the space to another level. Here, the owner not only can freely feel the cultivation from the inner wealth but also can taste the great pursuit and interests of life, which is intriguing.

the power of carnival

这套房子中，室内设计的亮点在于用来装饰的各种图案和摆设。设计师在纺织品、地毯和其他饰品上使用印花和重复的图案，营造不同寻常的视觉效果，强调个性化的气氛。在运用与众不同的模式的同时，设计师还善于运用色系搭配，通过对比设计构筑出一个有别于传统的空间，加上充分展示业主性格与喜好的独特家具，创造出一个和谐的突出一体化的室内环境。

多样的天然材料，以及公共区域里随处可见的丰富的绿色植物，带给人一种舒适感，在氤氲着宛如深沉的绅士风度的空间里点缀出活力与生气。卧室中的金猪托盘为空间添入一抹明亮的跳脱色，创意的黑白搭配又将空间质感提升到另一高度。在这里，可以肆意感受内心富足所散发出的涵养，也可以细品生活的大追求与小趣味，耐人寻味。

WHITE, AS SHAPE OF WATER

白色，如水的形状

Project Name | Apartment Panamby

Design Company | Maurício Karam

Designer | Maurício Karam

Project Location | São Paulo

Area | 250 ㎡

Photographer | Mariana Orsi

"Water is originally invisible, but can form a shape because of the pots holding it." While the white in colors is as same as the water. White is the most imaginative color like the invisible water, while when the designer designs in the white, space is given the emotion and personality. The design of this case with the theme of white injects the owner's personality into space and forms the shape that the designer wants.

The high-raised ceiling is the most eye-catching design of this project. In the design of the ceiling, the designer selects the simple technique to create a unified, bright visual effect without gorgeous decorations, which can satisfy people's pursuit of a simple, frank lifestyle at present. The walls and the inevitable bearing columns are outlined by the unified, concise lines, which increase the sense of comfort of the space yet not diminish the overall elegance and transparency.

The living room and the dining room adopt the open layout, and the decorations of furniture also adopt the black and white collocation, which cannot destroy the brisk aesthetic effect indoor. In the elegant fabric furniture, a leather orange and yellow multi-seat sofa light the whole design.

The dining room is on one side of the living room, and the elegant hanging chandelier shows the noble identity of the owner. A large old mirror is an artwork carefully designed by the designer. With the champagne light atmosphere, you can see the subtle expression of the guest and the elegant dining action of the owner.

The layout of the bedroom also continues the white-styled design technique. The designer tries not to destroy the integrity of white space as little as possible and uses two contemporary mirrored nightstands to add a fashionable temperament of the bedroom. While a passionate and connotative burgundy bed chair represents the owner's open personality.

The painting in the room is always disappearing and reappearing, continuing the story of the past. Architecture has its own experiences, while interior has its own stories. Therefore, the designer wants to combine the classic and modern as well as the tradition and fashion and presents the history and cultural connotation of interior space perfectly.

“水本无形，因器成之”，而白色在五光十色中也是如此。白，是最富想象力的颜色，如水一般无形无格，而当设计师在白色上落笔，便赋予了它情绪与性格。本案设计便是以白为主题，往空间里注入了业主的性格，形成了设计师想要的形状。

项目最为抢眼的是整体高挑的天花。在天花板的设计中，设计师选择用简约的手法打造整洁明亮的视觉效果，没有华丽的装饰正符合当下人们对简洁率真的生活方式的追求。墙体与不可避免的承重柱采用统一的简洁线条加以勾勒，反而增加了空间的舒适感，不减整体的文雅通透。

客厅与餐厅采用开放式格局，家具装饰一律采用黑白搭配，从而不破坏室内明快的审美效果。在一派典雅的布艺家具中，一张皮质的多人位橙黄色包扣沙发点亮了通篇设计。

餐厅位于客厅的一端，垂坠而下的典雅水晶灯彰显着居者的高贵身份，一面大型的做旧镜子是设计师精心挑选的艺术品，在香槟色的灯光氛围中仿佛能看到客人微妙的表情与主人优雅的进餐动作。

卧室的布置同样延续了白色派的设计手法。设计师尽量少地破坏白色空间的整体性，又用两个极具当代性的镜面床头柜增加卧室的时尚气质，一张热情而内涵的酒红色床尾凳代表着主人的开放个性。

室内的油画总是时隐时现，延续着过去的时代故事。建筑有自己的经历，而室内也有自己的心事，设计师则希望将古典与现代、传统与时尚兼容并蓄，把室内空间的历史与文化内涵完美地表达出来。

THE UNIQUE RESIDENCE · PHYSICS

型格 · 形而下

Project Name | WARREN ST. WRITEUP

Design Company | Ghislane Vinas Interior Design

Designer | Ghislane Vinas Interior Design

Project Location | Tribeca, New York

Photographer | Eric Laignel

Designers were able to repurpose a lot of the client's previous furniture in creative ways, and used tons of color to liven the space up. In the back of our minds, we had to remember that not only was this a home for small children but also a place for the client to entertain and to show off her art collection.

Starting with the entry, once passing through the massive black doors you're immediately struck by a custom designed wallpaper of oversized polka dots and a graphic scroll pattern of grays and oranges. Anchoring the space is a vintage orange table with a giant chandelier made from ping pong balls. To the left are a pair of vintage chairs and upholstered to look as if they were dipped in paint at different points and took an existing floor lamp and had it wrapped in cream crochet.

On the third floor you find the master suite and Paige's office. The suite is a respite from all the color surrounding the rest of the house, done in soft grays. The fourth floor is the kid's domain, with the kid's bedroom, play area and two guest rooms. The kid's room houses all 3 of the boys and features a wacky mural by artist Mark Mulroney. The size of the room offers lots of space for the boys to play and have sleepovers as they grow older, but also allows for the option to split the room into two if so desired later.

Next to the bedroom is the children's library, with low custom bookcases and beanbag chairs for easy reading. It also got red racing stripes, which carries out throughout the rest of the public space on this floor. Further down the hall you reach the kid's play area, with a large metal table for crafting, vintage Panton pendants above. Across from the play area are one of the two bathrooms on this floor, a custom corian vanity, and a tub with whimsical clouds made of neon above. Also, on this floor are two adjoining guest rooms, both in charcoal grays and bright greens.

The formal living room including dining room and kitchen is on the fifth floor. The living and dining rooms are decorated in black, with subtle vertical black stripes painted in differing finishes.

The dining room is using the existing dining table, but refinished in black with a bright yellow underside. The chairs were decorated with different plates to explore what would clients love to eat. On either side of the table are storage, the buffet with multiple legs and the taller one a vintage piece that was refinished in black on the exterior and bright yellow on the interior.

Over in the kitchen area, the colors were determined by a piece of art by Lisa Ruyter. The most spectacular part of the kitchen may very well be the bright yellow island, done with overly ornate molding, and housing hidden cabinets for storage.

The sixth floor houses the lounge area. The front space is the library and movie watching area, with a bright grass green rug and comfy chaises for cuddling. Both walls have the floor to ceiling book and movie storage in white oak. The back area is done in all white, with an incredibly comfortable sectional, a light up white coffee table and small white and oak side tables. This is the brightest area of the house because its wall to ceiling glass and overlooks a small deck area.

设计师以创新的方式重新设计客户以前的家具，并使用数以吨计的色彩染料使空间活跃起来，为空间注入了新的活力。在设计师的脑海中——这不仅是一个有小孩的家，也是供业主娱乐和展示收藏品的艺术空间。

从入口玄关开始，穿过一扇黑色大门，你会立刻被定制墙纸所打动——超大的圆点花纹和灰色、橙色有序排列的图案。长廊的亮点是一张复古的橙色桌子，顶上有一盏大型吊灯，那是用乒乓球加以设计制成的。左边是一对白色搭配橙色的复古椅子，装上软垫的椅子看起来像是注入了饱满的油漆，旁边摆一盏奶酪色织品包裹的落地灯。

三楼是主人套房和办公空间。主人套房以柔软的灰色为主，灰色能使人们从周围的其它亮色中得到缓冲。四楼是儿童乐园，有儿童房、娱乐区和两间客房。儿童房里可以容纳三个男孩，其中有一幅艺术家马克·马尔罗尼的古怪壁画。房间的大小可以为男孩们提供很多玩耍空间，日后也可以选择把房间一分为二。

卧室旁边是儿童书房，有矮的书柜和豆袋椅，便于阅读。红色赛车条纹连接了这层楼的其他公共空间。穿过娱乐区，则可以来到有着浴室的楼层，这里有定制的可丽耐梳妆台、浴缸和一盏云朵形状的霓虹灯。两间相连的客房也在这层楼，都采用了炭灰色和明亮的绿色作为主色调。

五楼是正式的客厅，包括有餐厅和厨房。客厅和餐厅都以黑色为主，用细腻的垂直黑色条纹漆成不同的装饰，墙上挂着三幅人像挂画。

餐厅使用业主已有的餐桌，但用黑色和亮黄色的底面重新油饰。这些椅子是对顾客喜爱的食物的探索，每把椅子上都印有一个不同的盘子，桌子的两边都是储藏柜，多桌腿自助餐桌和稍高的复古自助餐桌，其外表都是黑色的，内部是亮黄色的。

在厨房区域的设计，颜色的运用以丽莎·鲁特的艺术品为主导。厨房最壮观的部分可能是亮黄色的中岛台，有着过于华丽的造型，并内置隐藏橱柜作为储藏功能。

六楼是休息区。前厅是图书馆和电影观赏区，墙壁上的白橡木储藏柜用来存放书和电影，而一张明亮的青草地毯给人舒适的躺椅供人享受。因为有落地玻璃，所以这是房子最明亮的区域，另外有一个小的露天天台可以俯瞰远处的风光。

This is an apartment built in an artistic deco building in 1952, one of the best buildings in Brussels, with a mix styles of the early 20th century and the early 21st century. In the continuation of the architectural style, at the same time, the designer designs the interior decoration to be simple and rational, exude a trace of classical charm in the modern style, be rich of historical heritage characteristics, and highlight the human feelings of the space.

The overall color atmosphere of the space is neutral, and some bright colors, allocating with lighting and furniture from Scandinavia, Italy, Belgium, France, Spain, Afghanistan, etc. show creativity in serenity. Using the white as grounding, the arched door seems to divide the space into two areas, which have the coherent fishbone floor, same lamps and two same carpets in different colors, showing the independence and connection of these two areas. The designer uses various elements repeatedly to echo with the same color of different objects and present the same personalized furniture and design characteristics in different areas, which give a jumping unified aesthetic feeling.

OVERLAPPING ELEMENTS IN THE WHITE, THE COLORFUL DECORATIVE CHARM

白底下的元素叠加，色彩纷呈的装饰魔力

Project Name | Emile Duray Apartment

Design Company | Detrois S.A.

Designer | Michel Penneman

Project Location | Brussels

Photographer | Serge Anton

BUILDING
A NEW
MILLENNIUM

一幢建于1925年的艺术装饰风格大楼，是布鲁塞尔最好的建筑之一，而这套带有20世纪初和21世纪初混合气息的公寓，便是坐落于此。设计师在延续建筑本身风格的同时，将室内装修得极简、极具理性，在现代中发散出一丝古典韵味，富有历史的传承特色，突出空间里的人情味。

空间整体色彩氛围呈中性，附带一些明亮的色彩，搭配来自斯堪的纳维亚、意大利、比利时、法国、西班牙、阿富汗等地的照明和家具，在宁静中表现出十足的创意。以白色打底，拱形门仿佛辟出两块天地，连贯的鱼骨纹地板，一样的灯具，同款不同色的两块地毯，各个区域彼此独立，又各有联系。设计师别出心裁地重复使用多种元素，令不同物件中的相同色彩相互呼应，令不同区域出现相同的个性家具或设计特点，给人一种跳跃的统一美感。

SOL LEWITT A RETROSPECTIVE
PUTMAN STYLE
Sahara

MUERTE ERES

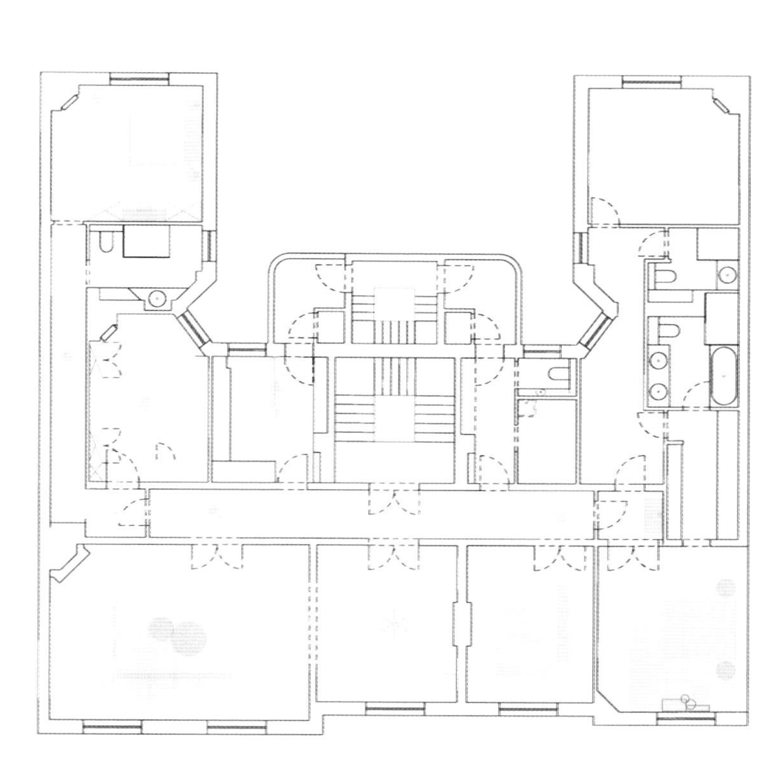

ENCOUNTER: BETWEEN THE COLORFUL AND PRISTINE FASHION

遇见：在浓墨重彩与古朴时尚间

Project Name | Foch Downtown

Design Company | Vick Vanlian

Designer | Vick Vanlian

Project Location | Downtown, Beirut, Lebanon

Area | 260 m²

Photographer | Vicky Moukbel

This house aims to create something different and beautiful, evoke emotions and present enough flexibility to let the home evolve in style with time.

The views of the Roman ruins outside are overpowering and phenomenal. While the solid oak wood floors and eggshell color walls inside form a soothing palette to provide the perfect backdrop for the owner's collections of interior pieces. At a certain moment, the sunshine outside the house comes in, and the colors in the room become more active. The designer also uses the contrasting furnishing ingeniously like the mirrored console and the bespoke high-gloss bar with jagged edges and takes advantage of materials and geometries to create a space that is very personalized and loyal to its owner's ideal.

Colors and patterns are chosen to match the moods of each room. The main living area is white, rendering a relaxing, inviting ambience. A bright patterned carpet and some vibrant art pieces including a Salvador Dali, punctuate the space delicately with color. Then, the Coca Cola panel is originally an old garage doors from the 70's, which if it could, it would tell an amazing story about the revolution of Beirut. In the master bedroom, a Versace metallic turquoise bed blends in perfectly with the whole room decor; a magnificent collage art work with golden frame by Manuela Crotti, forms the headboard. Space is colorful and full of artistic atmosphere. The designer seems to be brewing in a unique way, and then paints the owner's wonderful family story.

LOUIS VUITTON : 100 LEGENDARY TRUNKS

LIBAN
IAM

IAM

本案旨在创造与众不同的美好观感，唤醒人们对美的感受，同时展现出足够的灵活性，让整个家庭的风格随着时间的变化而变化。

室外的罗马废墟景观具有压倒性的非凡效果，而室内坚实的橡木地板和蛋壳色的墙壁形成了一个舒缓的调色板，为业主收藏的室内艺术品提供了完美的背景墙。某一时刻，屋外阳光照进来，屋内的各色便随之变得活跃。设计师还巧妙使用对比鲜明的饰面，如镜像台面、带有锯齿状边缘的定制高光泽吧台，利用材料和几何图形来创造一个个性十足但又忠于业主想法的空间。

B.B.M.
SIZE
IBx103

颜色和图案的选择与每个房间的心情相匹配。主要的生活区以白色为主，渲染一种轻松、诱人的气氛。一张明亮的带有图案的地毯，以及一些充满活力的如西班牙画家萨尔瓦多·达利所作的艺术作品，色彩微妙地点缀着空间。可口可乐面板原本是一扇20世纪70年代的老式车库门，如果可能的话，它将会讲述一个关于贝鲁特革命的惊人故事。卧室中，范思哲的金属绿松石床与整个房间的装饰完美融合，出自意大利设计师Manuela Crott的带有金框的宏伟拼贴艺术作品构成了床头板。空间里色彩斑斓，艺术气息浓厚，设计师似乎在用独特的方式酝酿着，而后描绘出业主的精彩家庭故事。

THE STREAM OF CONSCIOUSNESS OF DESIGN

设计意识流

Project Name | Captions Neuilly

Design Company | SC edition

Designer | Stephanie Coutas

Project Location | Paris

Area | 400 ㎡

This 400 square meters house is located in the beautiful Parisian suburb of Neuilly-sur-Seine, and has been stylishly crafted by the interior designer, Stephanie Coutas, who has created an extravagant space where contemporary art meets pleasant elegance. This two-storey villa has its own extensive 1000 square meters garden.

In this calm and quiet space, the challenge is to create an innovative and decorative interior, with one of the main features being the amazing range of contemporary art. A series of retro photographs, an artistic table football, a Plexiglas table by Peter Klasen, artwork by Robert Combas, a velvet gorilla and a sculpture of a fire extinguisher are featured in this house to create a refreshing and free arty style.

As with all of Stephanie's designs, this case fuses her love of beautiful workmanship and romanticism to create a neoclassical and contemporary space, imbued with modern magic. Much of the interiors are designed by Stephanie Coutas for 1001 Maisons and incorporate her signature style of working with high quality materials, such as marble and woodwork, to create an exclusive upscale resident space.

HYBER
Je vais

Whilst pursuing the high quality, the designer also envisions a beautiful family home. Stephanie uses shades of yellow, orange, bronze green and graphite black to freshen up the classic woodwork, which creates a warm and enthusiastic living atmosphere.

The function area in this space also has a kitchen, games room and two living rooms, all facing the garden. One of the living rooms has a fireplace, and the other has a study, which can be used as a drawing room or a working space.

With great knowledge and expertise, Stephanie shifts away from the conventional standards of a traditional home and creates a house full of fashion and entertainment that features an air of sophistication and leisure.

这个面积为400m²的房子坐落于塞纳河畔讷伊——美丽的巴黎郊区，由室内设计师斯蒂芬妮精心雕琢，创造出的一个奢华且具有当代艺术与舒适优雅的空间，此外，这个两层别墅还带有一个1000m²的大花园。

在这个沉稳和安静的空间里，创造一个具有创新性和装饰性的设计是非常具有挑战性的。这套住宅主要的特点之一是当代艺术的惊人范围。一系列的复古照片、一张极具艺术感的足球桌、一张Peter Klase设计的树脂玻璃桌、Robert Combas的艺术品、天鹅绒材质的大猩猩布偶和一个灭火器雕塑等等，这些都创造了一个令人耳目一新且自由随性的艺术风格。

与斯蒂芬妮所有的设计一样，本套设计融合了她对美丽工艺品和浪漫主义的热爱，从而创造出了一个融合新古典主义和当代风格的空间，充分展现出现代魔力。室内设计大部分是由斯蒂芬妮为1001 Maisons设计的，将她标志性的高质量材料融入其中，如大理石和木制品，力图打造独一无二的高档住宅空间。

在追求高品质的同时，设计师还憧憬着一个美好的家庭，斯蒂芬妮用黄色、橙色、青铜绿和石墨黑的色调粉刷已有的经典木制品，给家营造一种温暖热情的生活氛围。

空间中的功能区设置有厨房、游戏室和两个客厅，所有的空间都面向花园。其中一个客厅带复古壁炉，另一个客厅设计了书房，可以会客也可以充当工作室。

斯蒂芬妮用她渊博的知识和丰富的专业技术改变了传统家庭固有的标准，创造了一个充满时尚和生活乐趣的房子，生活氛围既成熟又休闲。

Tous
les
jours
Je vais
de mieux
en mieux
à tous
points
de vue

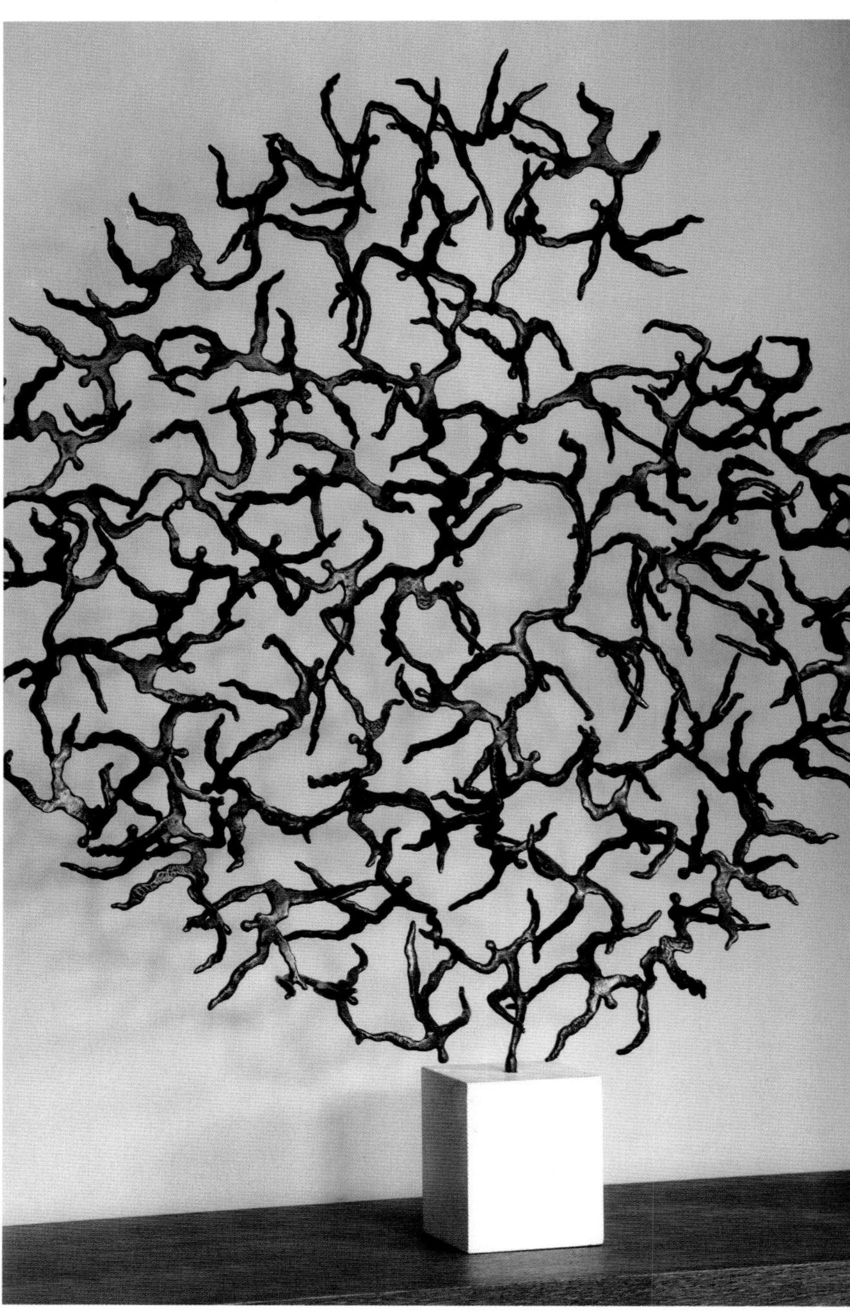

TRASH
Autopsies Kathy Reichs

AN URBAN, RETRO AND ABSTRACT RESIDENCE

Project Name I La Casa Bellezza

Design Company I Vick Vanlian

Designer I Vick Vanlian

Project Location I Beirut

Area I 250 m²

Photographer I Vicky Moukbel

都市尚古写意派

This apartment aims to express the unique and personalized story of the owner's journey. And the idea for this project is to combine vintage and urban to create a unique character that leaves an impression on the visitors.

The main entrance wall is painted with colorful colors shaping a girl's back and hair, which makes the static space have a dynamic beauty.

To emphasize the urban characteristic and vintage feeling of this house, the designer selects white bricks mixed with concrete to handle the wall's finishing, with a striking graffiti on it expressing the silhouette of Miles Davis and quoting Picasso's: "All you can imagine is real."

The exposed zinc pipes dominate the wall of the dinning room and are used as shelving. The pipes are actually reused pipes, as parts of the re-fuse collection, which aim to re-cycle old building materials for the purpose of minimizing waste.

Behind the dinning table, two Charbel Oun paintings became the focal point and contrasted with the black aim light from Bouroullec Brothers. The white and gray wooden dining chairs add a fashionable and tough temperament.

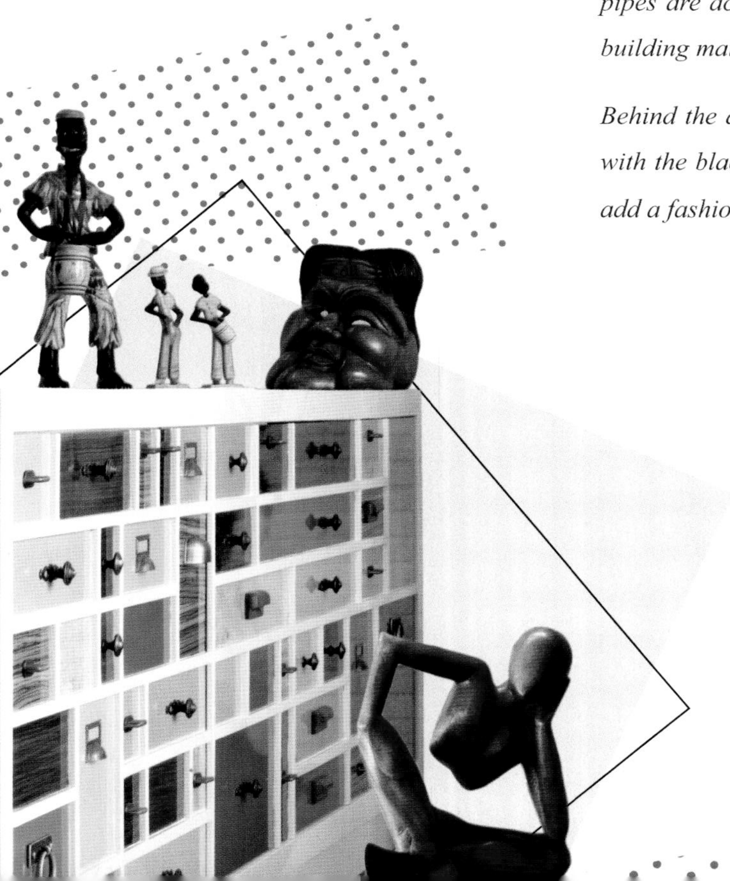

YOU CAN IMAGINE
ALL
IS
REAL!!!
PICASSO
DALÍ
TELEPHONE

YOU CAN
ALL
IMAGINE
IS
REAL!!!
PICASSO
PEINTURES
1955—1956
GALERIE LOUISE LEIRIS
47 Rue de Monceau—PARIS—VIII
MARS—AVRIL 1957
FESTIVAL
GLAMOUR
Nützliches Netz
ICE COLD
Coca-Cola

The London style phone booth is used to store the owners' motorbike helmets; behind the couch, the designer places three old vintage Coca-Cola bottle boxes as a display cabinet.

The designer uses a long cabinet with multi-drawer to separate the living and the indoor green area. The geometrical low tables in light turquoise color made of fiber glass and steel, a long couch and a chair form the indoor green area, which adds its unique rhythm of green leisure to the house with the exotic and eclectic feeling. At the same time, it hides the TV inside it, which aims to be able to watch TV from all areas in the house.

本案旨在表达主人在旅程中所经历的独一无二和饶有个性的故事，其设计理念是结合复古风和都市特点来创造一个独特的符号，看一次就能令你难以忘怀。

入门墙上用色彩勾勒出一个女孩的背部和飘扬头发，使得静态空间多了一份动态的美感。

为了强调这个房子带有的都市特点和复古感觉，设计师选用一种白色砖混合混凝土来处理墙的饰面，墙上一个引人注目的涂鸦显现出爵士乐大师——迈尔斯·戴维斯的轮廓，并引用毕加索的一句话："你所能想象到的一切都是真实的"。

裸露的锌管盘踞在餐厅的墙面，稍加改造便成为展示墙。这些管道是被重复利用而来的，也是重新整合的收藏品的一部分，它彰显着利用可循环的旧建筑材料来尽量减少浪费的环保精神。

餐桌后面挂着的两张Charbel Oun的画是视线焦点，并与Bouroullec Brothers的黑色射灯形成对比。白色和灰色相结合的木质餐椅增加了空间的时尚与铿锵气质。

英伦风的电话亭储物柜里存放着业主的机车头盔；而在沙发后面，设计师还摆放了三个做旧的印着"可口可乐"字样的箱子作为展示台。

设计师用一个多抽屉长柜来分隔生活区和室内绿化区。由玻璃纤维和钢制成的浅蓝色几何矮桌子、一张长椅和单椅子共同组成室内绿化区域，为富有异国情调和平衡感的房子增加了自身独有的绿色休闲感。同时，它可以将电视藏在里面，目的是让人能够在房子的各个区域看电视。

Club
ICE COLD
Coca-Cola
SOLD HERE

INTEGRATING CHINESE AND THE WEST, CONNECTING ANCIENT AND MODERN TIMES

Project Name | Puerta de Hierro

Design Company | AFTERL

Designer | Maria de la Osa

Project Location | Spain

Photographer | Natalia Apezetxea

融合中西　串联古今

The design of this case is permeated with the leisure pastoral life, and the casual home design creates an avant-garde aesthetic experience. The design derives from life, but also is applied and blended in life.

In the living room, the French window with a large area brings the natural light to space perfectly, and the overall white space gives people a neat, comfortable and elegant feeling. The multi-seat beige fabric sofa in a modern shape allocates with the emerald armchair in a retro shape to create a warm and leisure atmosphere of pastoral life. The bookshelves are designed on the walls of the living room as much as possible, and it can reflect that the owner is a well-read person. There are many classical paintings on the house which can highlight the cultural flavor and great knowledge of the owner.

The design of the small reception area is an American style. The animal ornaments on the walls as well as the chairs and carpet with animal pattern reflect a bold and generous charm of the owner's personality, full of masculinity. The design of courtyard is the typical American pastoral style, and the rattan weaving sofa gives people a sense of closeness.

ANIMAL
SWITZERLAND

SWITZERLAND

The designer designs an open dining room and kitchen. And the ring-shaped open window not only extends the sight but also makes the space bright and transparent. The red wallpaper printed the traditional Chinese painting reflects the different scenes of life in the ancient times of China. The cups, saucers and teapot of the western tableware allocating with the Chinese painting wallpaper form a different visual harmony. The knowledgeable owner always has a magnanimous temperament.

The design of the bedroom is concise and comfortable without any design language. The layout of the bedroom conforming to the life habits shows the designer's respect to the owner.

本案设计处处洋溢着闲适的田园生活气息，稍显随性的家居设计，却能营造出前卫的审美体验。所谓设计源于生活，而又寓于生活、融于生活。

客厅大面积的落地窗设计使得空间自然采光达到最佳，整体白色的空间环境，给人整洁舒雅的感觉。现代造型的多人位米色布艺沙发搭配复古造型的祖母绿单人沙发椅，营造了一个温馨闲适的田园生活氛围。客厅墙面设计了尽可能多的书架，从中可以看出居者是个博览群书的人，住宅中许多古典挂画更是突出居者的文化气息与丰富内涵。

CORN

小接待厅的设计则更具美式风味，墙面上的动物饰品，椅子及地毯的动物纹样，无不体现主人豪放大气的人格魅力，男子气概不言而喻。庭院设计是美式田园风的展现，藤编沙发给人以格外的亲近感。

设计师布置了开放格局的餐厅与厨房，厨房的环形开窗面不仅延伸了视线，还使得空间明亮通透。相互承继的红色壁纸上印着的是中国工笔画，一片片反映着中国古代不同的生活场景。西式餐具的杯碟茶壶混搭中国画壁纸，形成了不一样的视觉和谐，博闻广识的居者总有着海纳百川气质。

卧室的设计简洁舒适，没有过多的设计语言，顺应生活习惯的布置是设计师对居者的尊重。

Cincuenta sombras de Grey
E.L. James

THE RULELESS HOME FASHION

无规则家居时尚

Project Name | Kammakargatan 38

Design Company | Oscar Properties

Project Location | Stockholm, Sweden

Photographer | Henrik Nero

Main Materials | Fabric, Wooden Floor, Wood Veneer, Marble, etc

The design of a house cannot consider the styles and types, but without exception, home is comfortable and livable. The overall design of this case selects white as the base, and gray as an auxiliary color. On the basis of these, the designer places the owner's favorite furniture and ornaments. The whole space gives people a clean yet not lack of fashionable sense of beauty.

Walking into the door, you can feel a sense of friendliness. The multi-seat fabric sofa in a soft shape, the woolen beambag, and the enclosed layout of the living room give people a harmonious on-the-spot experience. The overall gray tone embellished with the bright red shows fashionable and lively. The wings spread in the hanging painting make the space full of tension, and its metal material shows a new aesthetic experience in a new century.

The dining room integrates with the kitchen, and the clean and neat space layout and furniture arrangement give people a fresh and comfortable space experience. The old wooden dining table, the dining chairs with a simple shape, the fashionable chandelier as well as the plants and flowers, all of these represent the owner's pursuit of life sentiment. Lighting several candles, space is full of a beautiful and romantic atmosphere.

The courtyard surrounded by the tall buildings is very comfortable. The design of this concise and generous courtyard can provide a comfortable place for the family to enjoy the afternoon. And the sunshine can embody the owner's heart of pursuing the genuineness in the bustling city.

The design of bedroom is leisurely and simple. The decoration of the headboard prefers to a Scandinavian style, which shows a dreamy naivete. The metal-clad bathtub is the highlight of the space, combining with a fashionable, transparent, ghost-shaped chair at the end of the bed to uniquely interpret an avant-garde aesthetic of the designer.

家可以不论风格，不讲门派，但无一例外都是舒适宜居的。本案设计整体以白色为底，灰色为辅，以此为基础摆设居者喜爱的家具、饰品，整体给人洁净大方又不失时尚的美感。

本案进门便感觉得到一股迎面扑来的亲近感，造型柔美的多人位沙发、毛线懒人沙发等，以及合围式的客厅布局都给人以融洽的临场体验。整体灰色调在少数明红色的点缀下显示出时尚与活泼，挂画中铺展开来的翅膀赋予了空间十足的张力，其金属材质给人新世纪新颖的审美体验。

餐厅与厨房相融合，干净利落的空间布局与家具布置给人清爽舒适的空间体验，做旧的木质长桌、造型简约的餐椅、时尚的吊灯，摆放的绿植鲜花是居者对生活情调的追求，点上几支蜡烛便洋溢出唯美浪漫气氛。

被高楼包裹着的小庭院显得分外宜人，简洁大方的庭院设计能给家人一个舒适的午后，阳光披洒下来体现的是居者在繁华都市里保有一份追求本真的心。

卧室设计最是休闲简易，床头装饰偏北欧风情，有几分童话般的率性纯真。金属包面的浴缸是空间的亮点，结合床尾一把时尚的透明鬼椅，是设计师前卫审美的独特演绎。

DEDUCING A PERFECT LIVING EXPERIENCE

极致演绎完美的居住体验

摄 影 师—江河摄影

主要材料—雅士白大理石、劳伦黑金大理石、水曲柳索色、墙布、黑镜钢等

项目面积 | 800 ㎡

项目地点—广东深圳

设 计 师—林志豪

设计公司—HCL设计事务所/深圳市林志豪装饰设计有限公司

项目名称—深圳莱蒙水榭山别墅

The place where every story happens has a kind of beauty. There, we can encounter with the beauty and be touched by the beauty. Then, the heart is touched, and the new story happens.

The living room is fashionable and avant-garde. The unique arc marble parquet and the noble metal glass surface form a unique visual feeling. A large area of high-end gray brings a rational feeling, while the sense of the rigid lines on the walls and ceiling forms a subtle inner relationship between the materials and techniques to constitute a layered and unified space imagination, like the beating notes making people feel comfortable. The sofa with high saturation of yellow is very striking, complementing to the lush plants, which is bright yet lively, rational yet sensible. And life is more comfortable. The beautiful life starts from the dining room, and the brisk hanging painting and the exquisite crystal chandelier make life more delicate. The decorations of easy-matching white, moderate gray and cool black bring the kitchen a unique beauty, which is concise and generous.

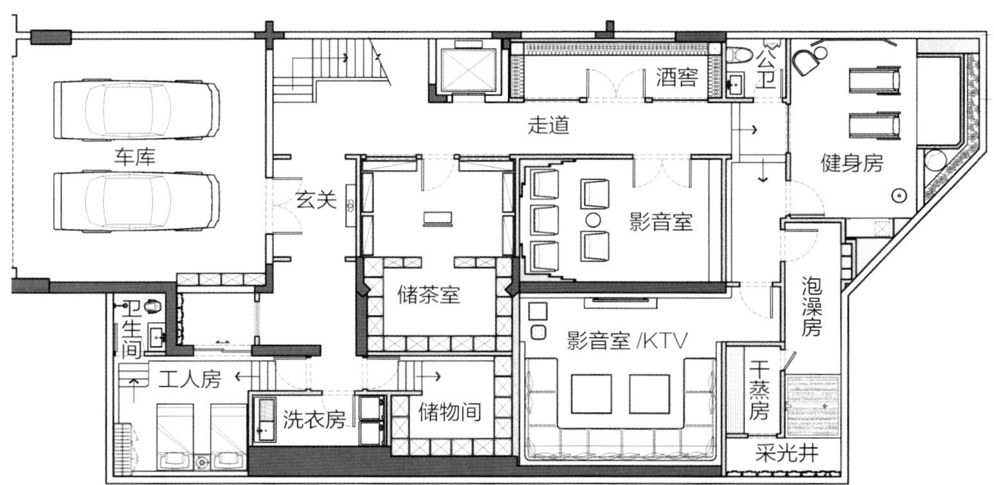

Walking up to the master's bedroom, the designer uses the personalized carpet with different shades of colors, the streamlined chandelier with exquisite technique, and the dazzling, elegant, fashionable suspended ceiling to show the elegant elements perfectly. The whole space brings people a broad horizon and a comfortable living experience from point to plane.

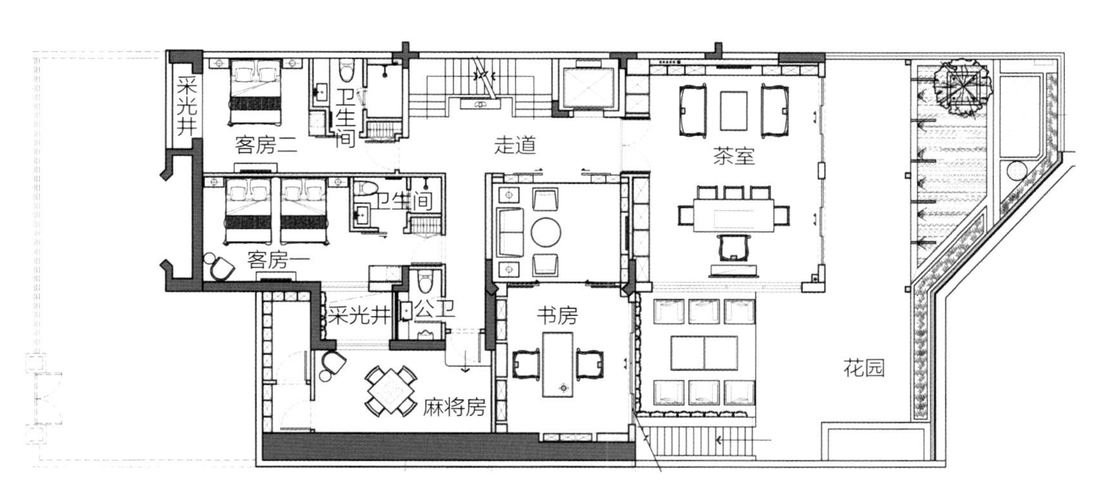
采光井
客房二
卫生间
走道
茶室
卫生间
客房一
采光井
公卫
书房
麻将房
花园

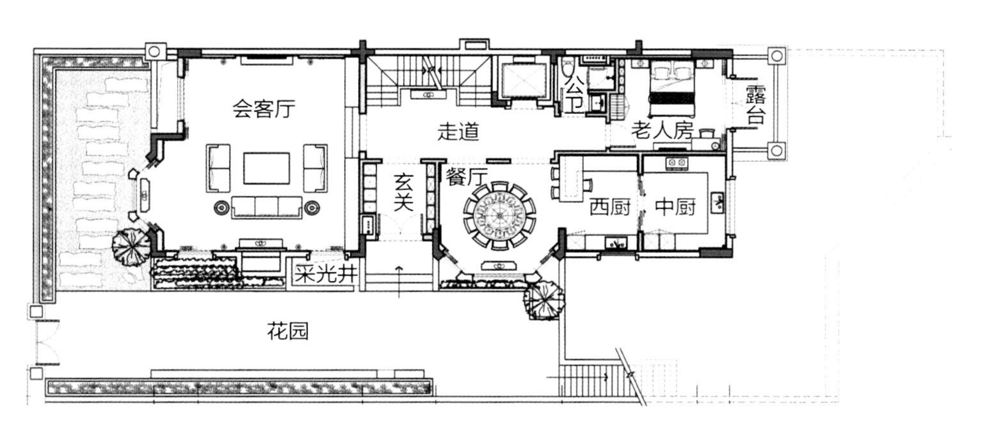

每段故事诞生的地方都有一种美的存在，与其邂逅，被其触动，心生涟漪，新的故事便随之诞生。

客厅空间时尚前卫，自成一格的圆弧形大理石拼花与高贵的金属玻璃面形成独特的视觉感受。大面积高级灰给人带来理性的气息，而墙面与天花立挺的线条感，在材质工艺之间结成微妙的内在关系，构成层次丰富而统一的空间想象，犹如跳动着的音符，让人心生惬意。高饱和度的黄色沙发极为醒目，与葱郁的植物相互补充，鲜明又活泼，理性又感性，生活更为轻松舒适。美好的生活从餐厅开始，一旁灵动的挂画与精致的水晶吊灯让生活更显精致。百搭白、中庸灰、炫酷黑的装扮，则让厨房有着独特的美感，简洁大气。

踏阶而上来到主人房，运用深浅不一的个性地毯，细腻工艺的流线型吊灯，璀璨高雅的时尚吊顶，将优雅元素体现得淋漓尽致，整个空间由点及面给人以开阔的视野和舒适的居住体验。

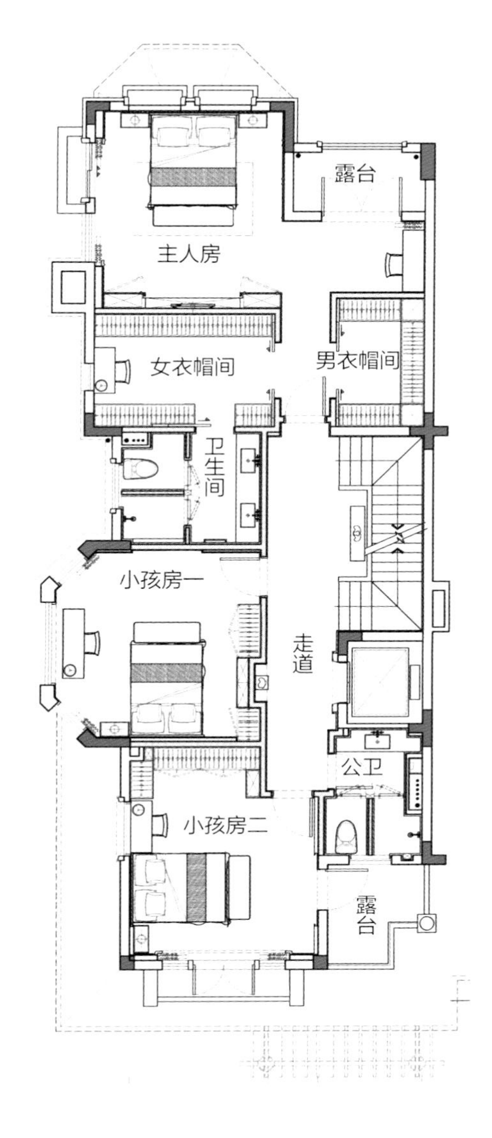
露台
主人房
女衣帽间
男衣帽间
卫生间
小孩房一
走道
公卫
小孩房二
露台

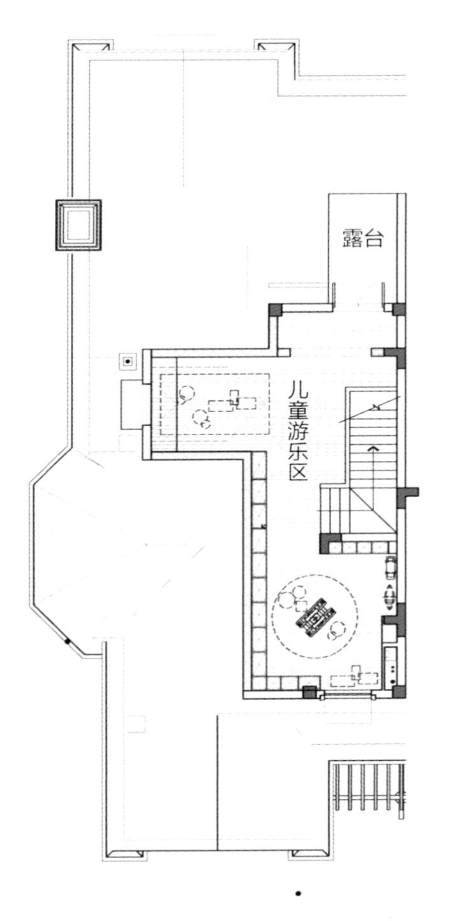
露台
儿童游乐区

FREE RANDOM THOUGHTS

自由随想录

摄影师 | 徐骁楚

项目面积 | 130 ㎡

协助设计 | 童妮娜、周胜杰、吴锋、高玲、潘维辰

总设计师 | 李想

设计公司 | 唯想国际

项目名称 | 唯想样板间

In the design of this model house, designers are unwilling to create an over-decorated and exquisite extravagance, but use the simple wallpaper paint, creative furniture forms without any furnishings and interesting allocations to show to us: life can be very free and easy, and the simple decorations can also show the elegance.

Designers want to create the most comfortable state at this moment freely, unrestrained by styles, and just hope it is natural, leisurely, comfortable and humorous. Do not worry about how to hide the ugly corners with repair, and whether there is an object for enjoying at the end of sight. The outlines on the walls painted by the brush just want to show a kind of insouciant beauty. The geometry-shaped furniture and the chairs with architectural hoods seem to be in an interesting negotiation.

We like the perception and calmness of the West, so that we turn the love to Rome into the table legs, and turn the feelings to Paris into the makeup of walls. We also are fond of the insouciance and leisure of the East, so that we put the alienation of mountain and the simple original wood color chairs there, like telling a kind of leisure.

PIANO

Mario
BOT
TA

The architect always unconsciously admires the sculptor's free creation. The long-admired rhythms sketched with the sculptor's brush are placed on the walls where everyone can notice. Life can be very casual, like the animal-shaped chairs, one is an honest elephant, and another is a smart cat. The artistic conception is hard to express, but, you can just express it freely according to your preference.

在设计这套样板房的时候，设计师不想创造浓妆艳抹、精雕细琢的贵气，只想用简单的壁纸、涂料、个性创意的家具，不用过多的陈设，镶嵌以趣味性的混搭来告诉我们：生活可以很洒脱，淡抹也可以很高雅。

设计师希望随心所欲地创造当下最舒服的状态。不讲求风格，只希望它是自然而然的，是闲适的、幽默的。不苦恼于如何用修补去隐藏丑陋的角落，不苦恼于视线的尽头是否有物守望。画笔涂出的墙上的线框只想诉说一种淡然优美，几何体的家具和带了建筑头套的椅子好似在进行有趣的谈判。

我们喜欢西方的感性深沉，所以把对罗马的爱恋变成坐榻的支脚，把对巴黎的情怀变成墙面的妆容。我们亦钟情于东方的淡然闲适，那么就把异化的山丘与原木色简洁的凳子放在那里，像在诉说一种悠然自得。

建筑师总是不自觉仰慕雕塑家的肆意刻画，用画笔勾勒那仰慕已久的韵律，就放置在目光会停留的墙边。生活本来就可以很随性，就像餐桌边的动物椅子，我本是憨厚的大象，你本是聪明的猫。意境难言，那就跟随喜好，自由表达。

A JOURNEY OF THE CONTEMPORARY VISION

一场当代视觉之旅

主要材料｜保加利亚灰石材、云多拉灰石材、黄铜金属、复合木地板等

项目面积｜170 ㎡

项目地点｜北京

设 计 师｜杜文彪

设计公司｜广州杜文彪装饰设计有限公司

项目名称｜北京万科滨江大都会170样板房

The male master of this case is a singer, who is attached to the music, and the hostess of this case is fond of art and design. Therefore, the design of this case becomes the leader in home art. Have an auditory journey, and find the contemporary Italian life here.

The ripples spread out on the wall of the entrance porch, transforming the invisible rhythm into a physical form and sketching the shapes that were born out of the feelings. The design and the music form an interfusion and symbiotic relationship with the wonderful traction.

Walking into the living room, there is an oncoming giant interesting hanging painting, allocating with the beige wall, like the spread of sound waves, as if in a world of sound. This is a custom artwork for this space by the design team, integrating the curves of a sound wave into the facial expression of people with the elegant, light colors, which can bring double feelings of nature and humanity.

The master bedroom abandons the complicated decorations of traditional Chinese style and makes the space boundaries clear. According to the main idea of the simple living style, the designer uses the transparent and bright lighting effects allocating with the interspersed artistic lights to make the whole atmosphere full of texture and permeability and create an extraordinary and luxurious experience.

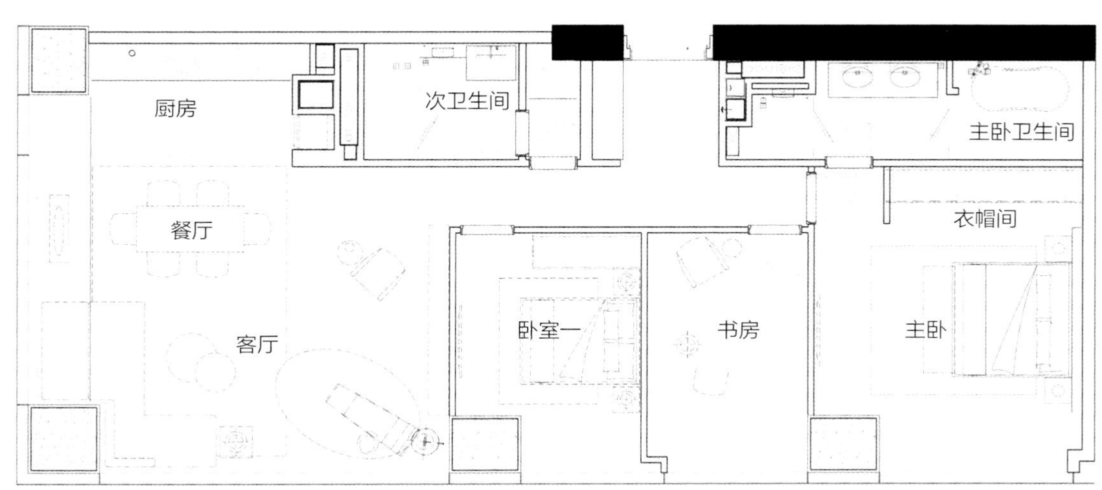

The boy's room is full of yacht elements, as well as the beacon-shaped lamp and the surfboard-shaped decorations are quite interesting and friendly. The high-end gray walls and the wood finishing in the dark color, allocating with the metal-textured lines reflect a decorative design of modern style, yet not lack of a sense of the quality of the design language.

本案设计中家居艺术的主导者，是寄情音乐的歌唱家男主人和爱好艺术与设计的女主人。来一场听觉之旅，寻找当代意式生活。

入户玄关墙上的涟漪散开，将无形节奏化为有形具象，并描摹出由情所孕育出来的形态，设计与音乐便在奇妙的牵引下相融共生。

步入客厅，迎面而来的是一幅灵趣的巨幅挂画，搭配米色的墙面，犹如声波的散开，仿佛置身于一个有声的世界。这是设计团队为空间特别定制的艺术品，将声波曲线融于人物的面部表情，在配以淡雅的色彩，带给人自然与人文的双重感受。

主人房隐去传统中式的繁复沉重，点到即止，明确空间界限，围绕简居简行的中心，通透敞亮的照明效果搭配艺术性灯光点缀，使整体氛围充满质感与通透性，打造出气质非凡的奢华体验。

男孩房是充满游艇元素的空间，灯塔形的吊灯，冲浪板形的装饰品颇具趣味感和亲切性。高级灰的墙面和深色木饰面，再加上金属质感的的线条，体现出现代风格的装饰性，同时又不失品质感的设计语言。

BEATLES

SILENT WORLD
ASSOULINE

ELITE STYLE: THE QUIET RESIDENCE IN THE FUTURE

摄 影 师－叶松、林峰
项目面积－500 ㎡
项目地点－浙江杭州
设 计 师－池陈平、袁杰
设计公司－尚层装饰（北京）有限公司杭州分公司
项目名称－世茂之西湖

精英气场：安静的未来居所

Since ancient times, the aesthetic consciousness of the Chinese cannot be formed without borrowing the scenery to express one's emotion and integrating emotion into the scenery. This is a kind of aesthetic interest established and connected by the subjective feelings and the "object", as an objective subject. In this case, designers bring us a different visual shock. The rough black and white marble walls and the exquisite white or black form a strong contrast, shape a sense of avant-garde and future, show the purest design spirit and create the soul of space. As designers believed, "For the contemporary design, pureness is not only a kind of style, but also a kind of artistic conception. Through the pureness of design, we can accurately express a spiritual realm or a living state of a certain will. And some designs seem to do nothing, but amazing."

“琴诗酒伴皆抛我，雪月花时最忆君”，自古中国人的审美意识便离不开借景抒情、融情于景，这是主观感情与作为客观主体的“物”建立关联的一种美的情趣。而在本案中，设计师给我们带来别样的视觉震撼。粗犷的黑白纹大理石墙面与细腻的白或黑形成强烈的对比，塑造空间的前卫感与未来感，展现极为纯粹的设计精神，铸就空间的灵魂。正如设计师认为：“对于当代设计来说，纯粹不仅仅是一种格调，更是一种意境，透过设计的纯粹性，可以准确地表达某种意志的精神境界或生存状态。有些设计，看起来似乎什么都没有做，但却是惊艳十足。”

The entrance porch with simple lines and concise facade seems to be empty. Only using the interleaving lights and the details of shell nosing as auxiliaries seem to bring people into the eternal design realm full of faith. The drawing room is full of masculinity yet not lack of fashion, like a dignified gentleman. A large French window frames the most beautiful scenery: the owner lives near the river with the surging river and the breaking waves out of the window, while the interior is quietness

and elegance. Less is more. Quietness and conciseness are reflected here incisively and vividly.

The dining area, consisting of 19 dining chairs and a stone dining table, has a powerful presence. All simple decorations are more noble and luxurious. Because of the reflection of natural lights, the glossy marble and walls divide the space layered. And the bright lights through the soft window gauze not only bring a little vitality to the atmosphere but also avoid the addition of too many colors. The owner can repast and have a meeting in this space. At the same time, through the open design, it connects with the living room, forming a transparent living pattern unconsciously which highly suits the aesthetic needs from a modern perspective.

Yu Qiuyu once wrote in The Agonized Journey of Culture, "Walks into the study, like walking into a long history, overlooking the vast world and cruising in the countless shining intelligent constellations. I become tininess, while becoming largeness suddenly. The study becomes a ritual, carrying the profit and loss expansion of life." Only if the home can give the homeowner the imagination, relaxation and the spiritual comfort, the dream of "home" will have the opportunity to come true. This is life, and it also makes designers plan the direction and purpose of the design to a certain extent, based on the preferences of the owner.

The bedroom emphasizes the power of quietness. Designers use a number of matte, leather, brown metal and fabric soft clothing, allocate with plain materials and integrate the international design techniques to emphasize the elite's tastes and create a comfortable and practical space, which can make the concept of "home" really return to the pursuit of a comfortable modern life. While the so-called luxury is the exquisite life and the quiet seclusion.

In addition to the master bedroom, this house also has four guest rooms. Designers mainly want to reflect the comfort and privacy, so that in the selection of materials, they choose leather, solid wood, fabric, carpet, and so on, and select the quiet gray and brown as matching colors to bring a comfortable and quiet atmosphere to the space. While the occasional lemon yellow and lake green inject a lively atmosphere for space like integrating the sensibility into the rational emotion, which makes space more balanced and comfortable easily.

入户玄关线条简单、立面干脆，看似空无一物，唯有光影交错和收口的细节作为辅助，似乎带人们进入充满信仰的永恒的设计境界。会客区充满阳刚之气亦不乏时尚，宛若一位气质轩昂的绅士；大面积的落地窗框出了最美的风景——临江而居，窗外江水滔滔、浪花朵朵，而室内静谧优雅。少即是多，安静、干练，在这里得到精确的诠释。

由19把餐椅和一气呵成的石材餐桌组成的就餐区，气场强大，一切装饰去繁从简，更显尊贵与奢华。光洁的大理石和墙面，因为自然光的反射，把空间分出层次，明亮的光线透过柔软的窗纱，让气氛带着些许灵动，也避免了过多色彩的加入。在这个区间里，可就餐、可会议，且开放式设计令其与客厅相连，于无形中形成通透的起居格局，高度符合现代视角下的审美需求。

余秋雨曾在《文化苦旅》中写道：“走进书房，就像走进了漫长的历史，鸟瞰着辽阔的世界，游弋于无数闪闪烁烁的智能星座之间。我突然变得琐小，又突然变得宏大，书房成了一个典仪，操持着生命的盈亏缩胀。”只有当家能够给予居者想象与放松、精神慰藉，关于“家”的梦想才会有成真的契机，这是生活，也是设计师基于业主的喜好，在一定程度上规划设计方向与目的的表现。

卧室重在表现安静的力量，设计师大量运用亚光面、皮革、茶色金属以及布艺软装，配合素色材质，融入国际化的设计手法，在强调精英品位的同时，更着力打造空间的舒适感与实用度，使“家”的概念真正回归到对舒适的现代化生活的追求中。而所谓的奢华，是极致的生活，也应是宁静的隐退。

除了主卧，本案还设置了四间客房，设计师以体现舒适和私密为主，材质上多选择真皮、实木、织物、地毯等，配色亦为宁静的灰色及棕色，为空间带来舒适安静的氛围。而偶然出现的柠檬黄及湖水绿，为空间注入活泼的气息，如把感性融入理性，并举重若轻，令空间更加平衡、惬意。

BARELY CALLED A VILLA

也是一墅

主要材料—大理石、木饰面、灰色烤漆、墙布硬包、清瓷水泥板等
项目面积—388 ㎡
项目地点—广东中山
设计团队—马英康、高颖颖、朱云锋、李永华
主案设计—彭征、谢泽坤、许淑炘
设计公司—广州共生形态工程设计有限公司
项目名称—丽丰棕榈彩虹花园别墅

This is a design project of the microcosm of a young age. The owner is curious about everything and wants to try everything. This is a dream age. Designers hope to design a living space by way of substituting for this plot to make it have the atmosphere of "habitat" and have the traces of people instead of being a simple show flat.

In the design process, firstly, designers make some minor modifications to the original plane of the building: the living room on the first floor is expanded; a large terrace on the third floor is intercepted a part to be a sun room; open up the basement and the garage, demolish workers room of the original building and move it to the innermost part of the basement to form a multi-functional room with a skylight, and the integrated parts are like the amber encasing a red Ducati "Diavel" motorcycle in a large glass cover becoming the focal point of the visual and spatial atmosphere and it has been transformed into an entertaining recreational space.

The design of the overall space is mainly in black and white, but the pure relationship of black and white will give the residential environment a cold feeling. Therefore, designers also intersperse the changeable yellow and gray color tone to make the space smooth and concise and give it a warm living atmosphere. On the one hand, through using the change of lighting and materials, designers create the subtle changes between the same color and the different colors. On the other hand, designers use the corresponding lines to cut the facade and the ceiling asymmetrically, integrate the lamps into space as the points and lines which are capable to activate space and give space a lively and interesting flavor.

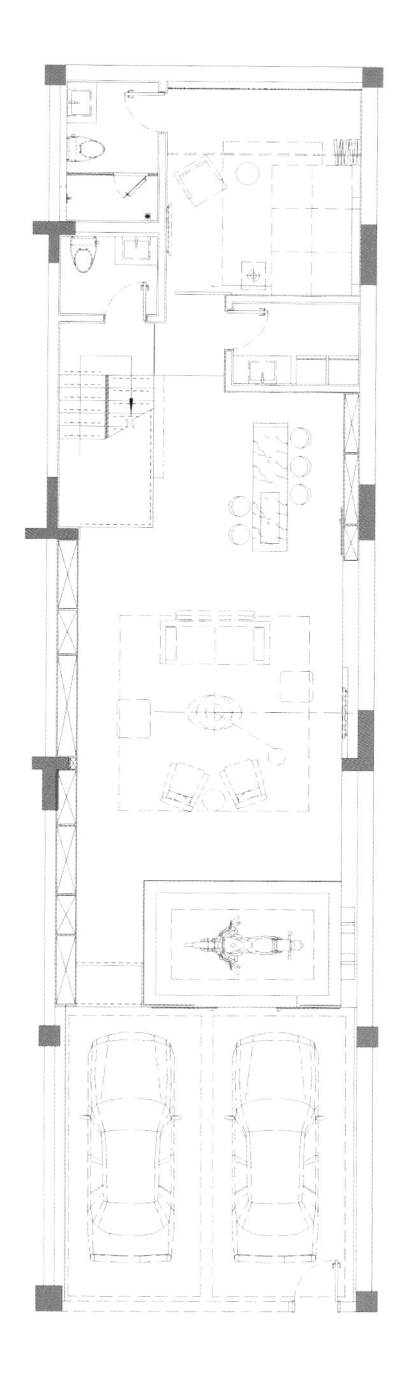

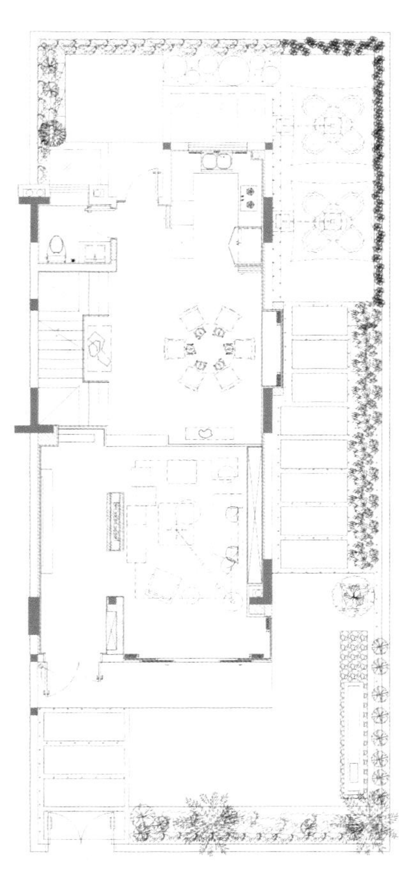

For the design of this project, designers do not regard it as a work to show off, but try to design it through thinking about the state of interior design. Taking account of the experiences of the use of designing the living environment, the characteristics of the building itself and other factors, designers finally achieve a relatively balanced result. The design of the living space is ultimately served for the owner. While the show flat, as a commercial display, is not only to sell the house but also a discussion of the possibilities of lifestyles—building a model in a hypothetical way, there will be the diversity of the lifestyle in this space in reality. Designers merely express one of these possibilities, but it may be based on the analysis of people and the state of life.

The 10Ks of Personal Branding
World wide internet
THE KISS
BLACK

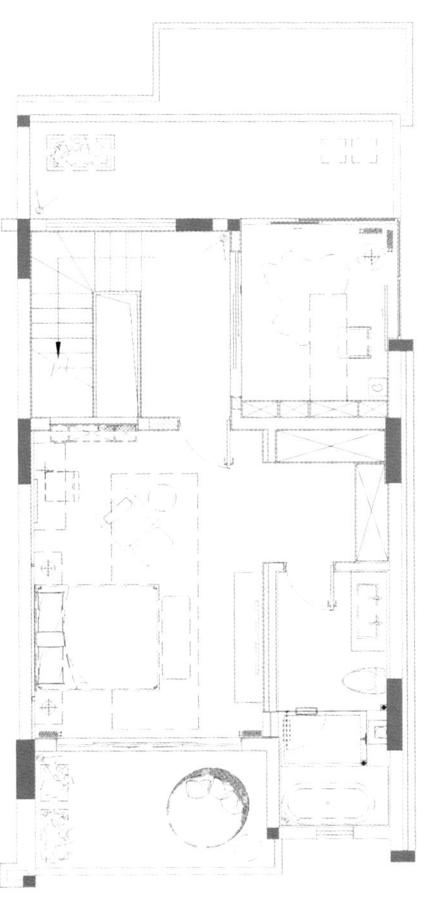

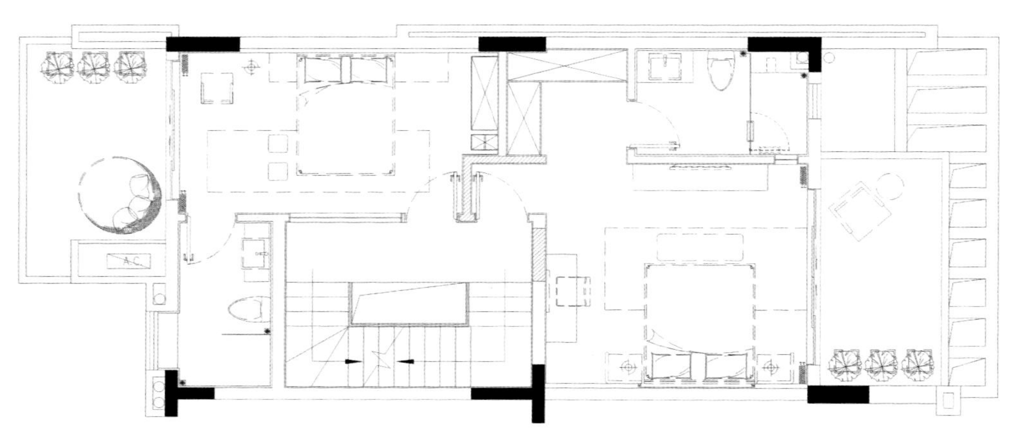

这是一个年轻时代的缩影的设计项目，主人对一切都好奇，对一切都想尝试，处于一个逐梦的年龄。设计师希望通过这种情节代入的方式来设计一个居住空间，使之完成后具有“人居”的气息，有人的痕迹，而不是一个单纯的样板房。

设计师在设计的过程中，首先对建筑的原平面作了一些小改造：一楼的客厅外扩了一部分；三层的大露台截取一部分做了一个阳光房；负一层与车库打通，将原建筑的工人房拆除，将之移至负一层最靠里的翼侧，形成一个带天窗的多功能房，而将整合后的部分，以如琥珀一般将一辆红色的Ducati“大鬼”包进大玻璃罩为视觉和空间氛围的重点，改造成一个带有娱乐性质的玩赏空间。

对于整体空间的设计，设计师以黑白调子为主，但是纯粹的黑白关系会使住宅环境给人以冷冰冰的感受，因此设计师同时将变幻的黄色和灰色调子作为穿插，既让空间流畅简洁，又赋予其温馨的居住气氛。一方面通过使用灯光和材质的变化营造同一色系和不同色系的微妙变化，一方面通过相应的线条将立面和天花作不对称切割，将灯具作为能够活泼空间关系的点和线融入到空间当中，也赋予空间活泼的趣味。

该项目的设计，设计师并没将它当作可以炫耀的作品去完成，而是试图通过对业态的思考，做针对性的设计。通过对居住环境设计使用的经验，建筑本身的特点等诸多因素综合考虑，最终取得一个相对平衡的结果。居住空间的设计，最终是为了使用者本人服务。而样板房作为一种商业性质的展示，不仅仅是为了销售房子本身，设计师通过它，更是对一种生活方式的可能性的讨论——在假想中建立一个模型，在现实中这个空间会存在居住的多样性。设计师仅仅只是表述其中一种可能，但这种可能是建立在对人和生活状态分析的基础之上。

WHEN I CAME BACK FROM THE WORLD

当我从世界回来

项目面积 | 560 ㎡
项目地点 — 浙江杭州
软装设计师 — 陈燕
主创设计师 — 谢军
设计公司 — 尚层装饰（北京）有限公司杭州分公司
项目名称 — 融科瑷骊山

The owners of this case are a couple who were born in the 1990s, and they advocate aesthetics. Through self-employed, they create their fashion kingdom from nothing. Because of the needs of work, they have been to many countries for shooting. The broad horizon and the keen sense of fashion make them have a higher demanding of the design. They don't like the complicated stacks and tough decorations and want to present the exquisite cultural design elements that they have seen from all over the world in their houses.

Designers integrate the modern style with the exquisite art ingeniously. The eternal neutral tone is harmonious with the simple materials with abundant textures. The fabric with various materials and the metal-textured furniture reconcile a natural, comfortable, elegant and exquisite style of the modern life.

The living room that is a public space with an open pattern presents luxury and generousness; the mosaic floor blends with the ivory white wall decorations skillfully; the warm brown color tone like the white chocolate flows in this whole white space, which is warm and textured.

The large-scale sofa is comfortable and textured, allocating with the interesting chair and the single-seat sofa with marble-textured fabrics to show a surprising harmony. The use of the geometry-textured fabric is to enrich the entire space.

本案的业主，是一对崇尚美学的90后，自主创业、白手起家，打造了属于自己的时装王国。因为工作需要，他们为拍片去过众多国家，开阔的眼界、敏锐的时尚嗅觉，都使他们对设计有了更高的要求。他们不爱繁沉复杂的堆砌，也不喜欢太过硬朗的装饰，希望能把他们从世界各地采风看到的精致且带有文化底蕴的设计元素呈现在房屋里。

设计师巧妙地把现代风格和精致的艺术结合在一起。永恒的中性色调，与简单却质感丰富的材质和谐并置，各种材质的布料和带有金属质感的家具，调和出一种自然舒适又典雅精致的摩登生活格调。

客厅这一公共空间以开放式的格局表现出奢华与大气，拼花地板与象牙白墙装饰巧妙地融合在一起，暖褐色基调像是整个纯白色空间流淌的白色巧克力，温润有质感。

大体量的沙发搭配怪趣张扬的单椅和大理石质感面料的单人位沙发，恰恰表现出出人意料的和谐。几何质感的布艺运用更是让整个空间丰富起来。

The most important design is the hanging plate decoration, FORNASETTI, on the walls, which is the inspired muse talked about by the designers. The angelic face and the perfect use of black, white and gold not only make the wall decorations structured and brisk but also ingeniously integrate the various elements with the colors in the space imperceptibly, which becomes a highlight.

Another emotional thread in the public area is the dining room and the social kitchen. This is a four-dimensional space sketched by designers intentionally to blur the boundaries of the area and make the scenes of life not only pass by but also stack on top of each other. The dining room extends the decorations of the living room well, but it is totally different from the living room. The black retro chandelier and the white antler lamp in the living room form a sharp contrast. And the FORNASETTI decoration is placed wittily in a cabinet full of European lines, which shows a surprising harmony and interest. Even in the dining room, there is still full of modern artistic flavour.

In the western kitchen, the cement-textured tall cupboard, transparent bar chairs, the wooden dining table, the velvet black dining chairs and the decorative cabinet with the colors and lines different from the surrounding decorations, all of these individual elements seem to be irrelevant and even incompatible. But together with them, they are seemly very harmonious as well as make people can not help praising the designers' taste and ingenuity.

The bedroom is the owner's private space, and the continuous use of the same color makes the whole space comfortable and warm. The use of gray linen bed and the geometric bed, as well as fabrics, fills up the loss of color tone in the space. The white chandelier adds a few dramatic charm to the whole space. The reading light and the metal-textured working area present a sense of moderness and art perfectly.

A lot of natural light fills the whole space, and the ingenious design integrates this seemingly outdoor recreation area full of vitality and greenery with the indoor drawing room. The metallic texture is used in this space maximally, full of freedom and unruliness of the young. The use of light gray handmade art paint on the walls and the artistic marble tiles on the floor shows the original textures and colors of nature. The whole space exudes the warmth of the wooden furniture, which can make the people in this space stay away from the city's hustle and bustle without any burden, and fully enjoy the harmony between people and nature.

Whether the Italy with a long history, having the ancient Roman culture yet not losing its splendor and elegance, or the lingering France combining with tradition and modernness, romance and shyness, or America with a blend of the culture around the world and even the air full of casualness and ease, the places where we had been together, the warm breath we had breathed together, and the beauties that are deep in our memory are brought back to our "home".

不得不提的是墙面的挂盘装饰——FORNASETTI，这个被设计界津津乐道的灵感缪斯，天使般的脸庞，黑、白、金色的完美运用，使墙面装饰规整而有灵性，又能在潜移默化中把空间里的各个元素材质及色彩巧妙地融合在一起，画龙点睛。

餐厅与社交厨房是公共区域中的另一条情感脉络，这是设计师刻意勾画的四维空间，模糊区域的界限，使得生活中的场景不只是擦肩而过，而是互相融合。餐厅空间很好地延伸了客厅装饰，却又完全不同，黑色的复古吊灯和客厅的白色鹿角灯形成了鲜明的反差，FORNASETTI装饰被俏皮地放进了一个充满欧式线条的柜子里，但却出奇的和谐有趣，即使是餐厅空间仍然充满了摩登的艺术气息。

西厨的水泥质感高柜、透明的吧椅、木质餐桌、天鹅绒面黑色餐椅，以及连色彩和线条都与周遭不一样的装饰柜，这些单独元素看似毫无关联，甚至有些格格不入，但放在一起却又不失和谐，让人不禁感叹设计师的慧眼与巧妙匠心。

卧室是主人的私享空间，相同色彩的延续运用，让整个空间看上去舒适温馨。麻灰色的床与几何床品及布艺的运用，填补了空间里的色调缺失。白色吊灯更为整个空间增添了几分戏剧性魅力。阅读灯及带有金属质感的工作区域，更是把摩登和艺术质感呈现得淋漓尽致。

大量的自然光线填充了整个空间，巧妙的设计使这个看似户外的充满生机绿意的休闲区与室内的会客区域融合在一起。金属质感在这个空间被最大限度地利用贯穿，充满了年轻人的自由和不羁。墙面浅灰色的手工艺术漆与地面大理石艺术砖的运用，呈现出大自然的原始纹理与色泽。整个空间将木质家具的温暖感烘托出来，让置身于此的人可以无负担地远离都市喧嚣，充分享受人与自然的和谐。

无论是历史悠久坐拥古罗马文化却又不失华丽典雅的意大利，亦或是缠绵缱绻，集传统与现代、浪漫与羞涩于一身的法国，还是融合了世界各地的文化，连空气都充满了随性轻松的美国，那些我们一起牵手走过的地方，那些我们一起呼吸过的温暖的气息，那些深存在我们记忆里的美好，就这样被我们带回了“家”。

CREATED FOR THE FUTURE, CLOSE TO THE WARM AND FREE LIFE

为未来而设，亲近温情自由生活

主要材料丨石材、砖、木作、乳胶漆、定制墙画等

项目面积丨538 ㎡

项目地点丨四川成都

陈设设计丨冯媛姣

主案设计丨郑军

设计公司丨耕图设计

项目名称丨成都复地雍湖湾城市联排别墅

There are unlimited possibilities in life, and one's preference cannot be judged by style rudely. Therefore, designers will not use "European", "French" or "modern" to outline a house, but through the owner himself/herself, the house can form its style. In this case, designers balance all kinds of free expressions and integrate art and modern characteristics. The reconstruction of space seemly cluttered but ordered presents that the owner has a distinctive understanding of life and reflects designers' thinking of lifestyle and the quality of life in the future.

Designers move the stairs in the interior space to outside to form an entrance porch, and take advantage of the reconstruction to fully strengthen the rationality of space. From the moving lines and structure, it seems to be injected more humanistic care. At the same time, the changes of pattern make the space more transparent and brighter, fade the atmosphere expressions of decorative elements, give the space free and highlight that the flow of light and air is space itself.

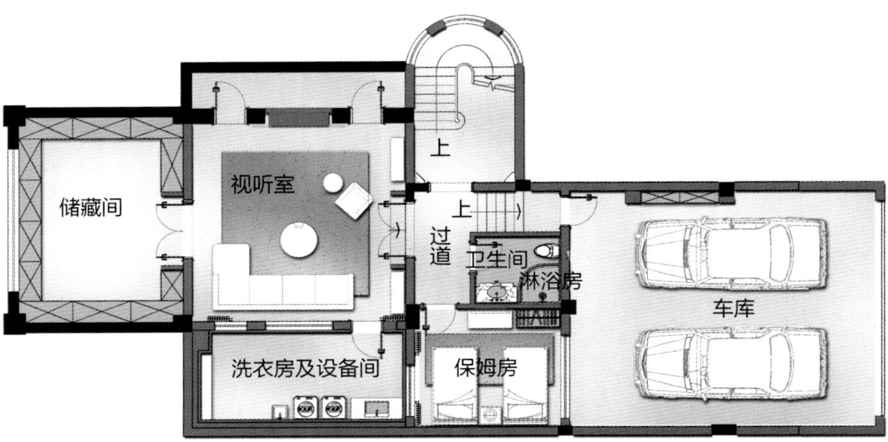

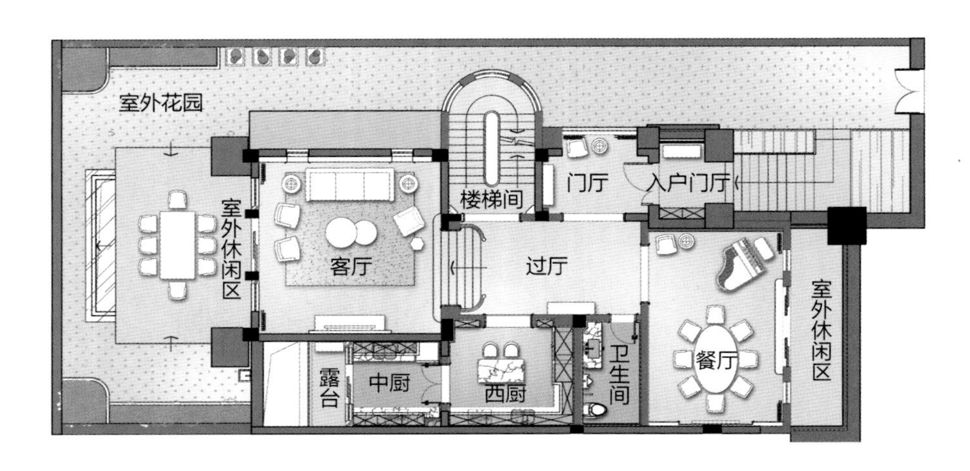

Geometry, sculpture, Fauvism, Impressionism, etc, the collision and the contrast among them can be seen everywhere in the space. It has not too many inductions of styles and also does not deal with the problems of more and less deliberately. The whole space creates a white surface effect deliberately which is convenient to accommodate more ego and qualities and lets the spatial temperament grow freely. Designers are also willing to build the scene of life, making both of the window-like brightness and the daily warmth in the space. The "retro" wall lines and the "modern" furnishing form a uniquely artistic charm. While the existence of the sunlight also covers a pleasant halo of space and highlights that the exquisite objects are not only the decorations but also the fun of life. It is particularly worth mentioning that designers have an interesting attempt to the colors and lines in the video room and hope that it is not usual, but changing naturally by means of the weirdly colorful film.

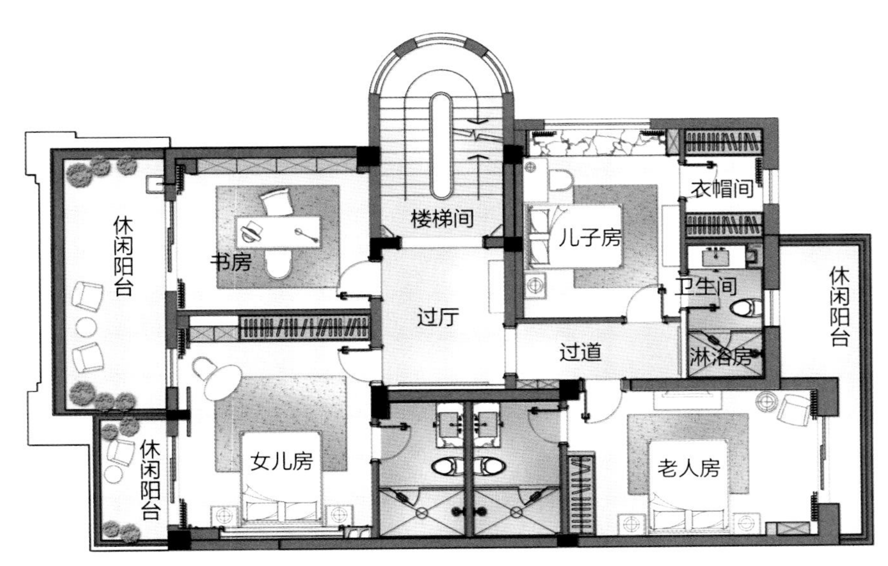

生活有无限多的可能性，喜好也不能以风格为标准进行粗暴的评判。因而设计师从不用“欧式”“法式”或“现代”来概述一个家，而是通过业主本身，形成自有的气场。在本案中，设计师平衡各种自由的表达，融入艺术和现代特色，对空间作看似无章实则有序的重构，呈现出业主对生活与众不同的理解，体现出设计师对未来的生活方式、生活品质的思考。

设计师把室内空间的楼梯外移，形成入口门廊，利用改建充分强化了空间的合理性，从动线和结构来看都更加注入人文关怀。同时，格局的改变让空间更加通透敞亮，褪去装饰元素的氛围表达，赋予空间自由，突出光和空气的流动才是空间本身。

几何、雕塑、野兽派、印象派……碰撞和对比在空间中无处不在，没有过多风格的归纳，也没有刻意处理多与少的问题，整体营造出白面的效果，方便容纳更多的自我和特质，让空间气质随性生长。设计师还乐于构筑生活的场景，使得橱窗般的光鲜和日常的温情片段在空间里兼而有之。墙板线条的“复古”与陈设上的“当今”，形成了特有的艺术韵味。而阳光的介入又为空间镀上惬意的光晕，更加突出精美的器物不止是装饰，更是生活的乐趣。尤为值得一提的是，设计师在影音室的色块和线条中进行了有趣的尝试，借由电影的陆离，希望它不似平常，又在情理之中变化。

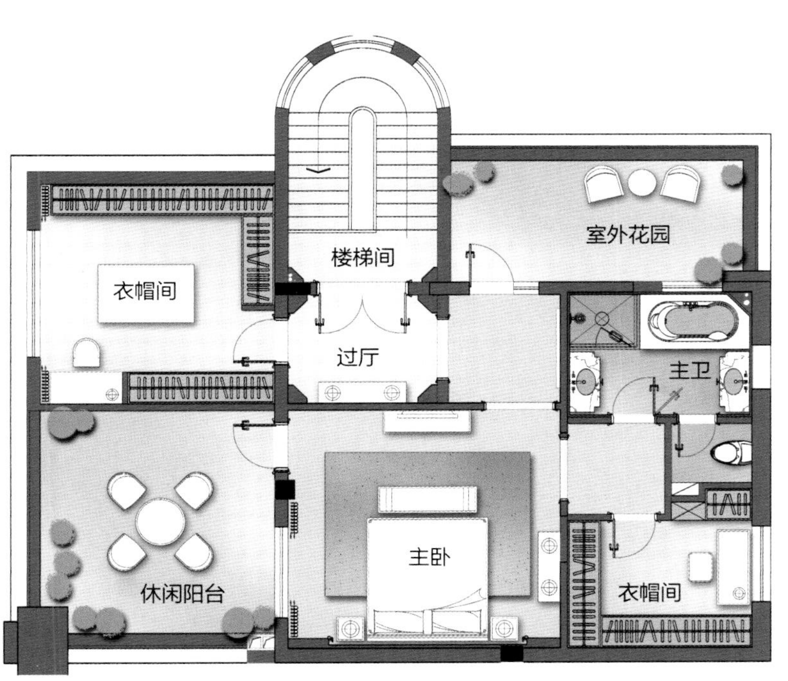

As Wang Xiaobo said, "In addition to a real life, people should have a poetic world." Art is another dimension of life, and the uninhibited and romantic heart is undoubtedly precious. Designers are moved by the subtle touches and stunning by the vigorous imaginations without any definition. In the space, the free combinations and storage seem to remind that we not only have a good appearance but also do not miss an interesting soul.

就如王小波所说，“除了一个现实的此生之外，人们还应该拥有一个诗意的世界”。艺术就是生活的另一个维度，不羁又浪漫的心无疑是宝贵的。令居者动容于细微的触动，惊艳于蓬勃的想象，设计师不落于任何定义，在空间中自由地组合与收纳，就仿佛提醒我们既要有姣好的面容，也不应错过有趣的灵魂。

图书在版编目（ＣＩＰ）数据

后现代美学设计 : 国际创新居住空间赏析 / 深圳视界文化传播有限公司编. -- 北京 : 中国林业出版社，2018.3
ISBN 978-7-5038-9457-2

Ⅰ. ①后… Ⅱ. ①深… Ⅲ. ①住宅—室内装饰设计 Ⅳ. ① TU241

中国版本图书馆 CIP 数据核字 (2018) 第 041755 号

编委会成员名单
策划制作：深圳视界文化传播有限公司（www.dvip-sz.com）
总 策 划：万 晶
编　　辑：杨珍琼
校　　对：陈劳平 尹丽斯
翻　　译：侯佳珍
装帧设计：叶一斌
联系电话：0755-82834960

中国林业出版社 · 建筑分社
策　　划：纪 亮
责任编辑：纪 亮 王思源

出版：中国林业出版社
（100009 北京西城区德内大街刘海胡同 7 号）
http://lycb.forestry.gov.cn/
电话：（010）8314 3518
发行：中国林业出版社
印刷：深圳市雅仕达印务有限公司
版次：2018 年 3 月第 1 版
印次：2018 年 3 月第 1 次
开本：235mm×335mm，1/16
印张：20
字数：300 千字
定价：428.00 元 (USD 86.00)